GRIT
AND
GOLD

Wilbur S. Shepperson Series in Nevada History

Series Editor: Michael Green (UNLV)

Nevada is known politically as a swing state and culturally as a swinging state. Politically, its electoral votes have gone to the winning presidential candidate in all but one election since 1912 (it missed in 1976). Its geographic location in the Sunbelt; an ethnically diverse, heavily urban, and fast-growing population; and an economy based on tourism and mining make it a laboratory for understanding the growth and development of postwar America and postindustrial society. Culturally, Nevada has been associated with legal gambling, easy divorce, and social permissiveness. Yet the state also exemplifies conflicts between image and reality: it is also a conservative state yet depends heavily on the federal government; its gaming regulatory system is the envy of the world but resulted from long and difficult experience with organized crime; its bright lights often obscure the role of organized religion in Nevada affairs. To some who have emphasized the impact of globalization and celebrated or deplored changing moral standards, Nevada reflects America and the world; to others, it affects them.

This series is named in honor of one of the state's most distinguished historians, author of numerous books on the state's immigrants and cultural development, a longtime educator, and an advocate for history and the humanities. The series welcomes manuscripts on any and all aspects of Nevada that offer insight into how the state has developed and how its development has been connected to the region, the nation, and the world.

A Great Basin Mosaic: The Cultures of Rural Nevada
JAMES W. HULSE

The Baneberry Disaster: A Generation of Atomic Fallout
LARRY C. JOHNS and ALAN R. JOHNS

History of Occupational Health and Safety: From 1905 to the Present
MICHELLE FOLLETTE TURK

Grit and Gold: The Death Valley Jayhawkers of 1849
JEAN JOHNSON

GRIT
AND
GOLD

THE DEATH VALLEY
JAYHAWKERS OF 1849

JEAN JOHNSON

UNIVERSITY OF NEVADA PRESS *Reno & Las Vegas*

University of Nevada Press | Reno, Nevada 89557 USA
www.unpress.nevada.edu
Copyright © 2018 by University of Nevada Press
All rights reserved
All photographs by LeRoy Johnson unless otherwise noted.
Cover photographs © by Doug Lemke/Shutterstock.com and LeRoy Johnson
Cover design by David Ter-Avanesyan/Ter33Design

LIBRARY OF CONGRESS CATALOGING-IN-PUBLICATION DATA
Names: Johnson, Jean, 1937- author.
Title: Grit and gold : the Death Valley Jayhawkers of 1849 / Jean Johnson.
Description: Reno ; Las Vegas : University of Nevada Press, [2018] | Includes bibliographical
    references and index. |
Identifiers: LCCN 2018011417 (print) | LCCN 2018013152 (e-book) | ISBN 978-1-943859-77-1 (pbk. :
    alk. paper) | ISBN 978-1-943859-78-8 (e-book)
Subjects: LCSH: Jayhawker Party. | Overland journeys to the Pacific. | Frontier and pioneer life—
    Death Valley (Calif. and Nev.) | Death Valley (Calif. and Nev.)—History. | Pioneers—Death Valley
    (Calif. and Nev.)—History. | Gold mines and mining—Death Valley (Calif. and Nev.)—History.
Classification: LCC F868.D2 (e-book) | LCC F868.D2 J6254 2018 (print) | DDC 979.4/87—dc23
LC record available at https://lccn.loc.gov/2018011417

FIRST PRINTING

Manufactured in the United States of America

To
LeRoy Johnson
without whom this book would not exist

CONTENTS

LIST OF ILLUSTRATIONS

Maps are by Eureka Cartography, Berkeley, based on the author's trail analysis. Photographs are by LeRoy Johnson unless otherwise noted.

Maps

Portraits

George Allen, photCL 1 (25)

Edward F. Bartholomew, photCL 1 (14)

Bruin Byrum, photCL 1 (17)

Juliet, John, James Brier, photCL 1 (21)

Charles Clarke, photCL 1 (18)

Alonzo C. Clay, photCL 1 (3)

John B. Colton, October 1850, photCL 1 (1)

John B. Colton, early 1850s, photDAG 5

John B. Colton, photCL 1 (2)

Urban P. Davison, photCL 1 (13)

Edward Doty

Edward Doty, photCL 1 (7)

Sidney P. Edgerton, photCL 1 (15)

Harrison B. Frans, 18 photCL 1 (10)

John Groscup, photCL 1 (16)

Captain Asa Haynes, photCL 1 (6)

Thomas McGrew, photCL 1 (24)

Charles B. Mecum, photCL 1 (5)

Alex Palmer, photCL 1 (12)

John Plummer, photCL 1 (20)

Figures

PORTRAITS

John Colton requested the Jayhawkers send photos for his collection and to share at the Jayhawker reunions. Original Jayhawkers for whom Colton had no photos were: W. Carter, John Cole, Marshall G. Edgerton, Aaron Larkin, and John (or N. D.) Morse. Men who were considered Death Valley Jayhawkers by the end of the trip but for whom I have no photos are Father Fish, the Frenchman, John Goller and his companion John Graff, Mr. Gould, Frederick A. Gritzner, William Isham, William Robinson, and Wolfgang Tauber (who traveled with Gritzner and Young). Colton had copies of the photographs made so he could return the originals if requested. Several originals were daguerreotypes; others were portraits taken many years before and were thus scratched or faded. (The photographer touched up hair and eyes on some photos and did considerable artwork on a couple others.) I used Photoshop only to remove dust specks and a few scratches, to crop photos, and improve contrast. When Colton knew the date of the original picture, he added it to the frame folder and often added the death date as well. I include photo dates where pertinent. Most of the portraits came into Colton's possession about 1893 when the Jayhawkers were already old men. Detail of portraits are courtesy of the Jayhawkers of '49 Collection, the Huntington Library, San Marino, California, unless otherwise credited.

George Allen, 1866

Edward F. Bartholomew

Bruin Byrum

Juliet, John, and James Brier

Charles Clarke

Alonzo C. Clay

John B. Colton, October 1850

John B. Colton, early 1850s

John B. Colton

Urban P. Davison

Edward Doty 1871

Edward Doty

Sidney P. Edgerton

Harrison B. Frans

John Groscup

Captain Asa Haynes

Thomas McGrew, 1860s

Charles B. Mecum, 1875

Alex Palmer, 1849

John Plummer

Luther A. Richards

William B. Rude, 1860s

Thomas Shannon

L. Dow Stephens

John L. West

Leander Woolsey

Sheldon Young, 1855

William Lewis Manly

Adonijah S. Welch

INTRODUCTION

Time can never erase from my memory
the close ties that bound us together in that trying time,
when from day to day we did not know who would be the next to say
to his comrades, "I can go no further."

—Death Valley Jayhawker Lorenzo Dow Stephens, 1899

"WE NEEDED NO MIRROR to tell us we were handsome—*we just knew it*" (Mecum 1878). The Jayhawkers of 1849 were carefree, jovial, self-assured, young men from Knox Country, western Illinois, who banded together as comrades on a quest for treasure in the newly discovered golden streams of California's Sierra foothills. Most were in their early twenties, and several had gone to school together (Colton 1897). They wore buckskins and took Indian-related names, and some shaved their heads in the style of the Otoe Indians (today called a Mohawk cut). Their initiation rites were held in the Platte River valley—Indian country—today's Nebraska, so I call this first brotherhood the Platte Valley Jayhawkers. During their adventures west, they forged bonds of friendship that later saved the lives of several of them as they struggled through the Nevada-California deserts. These men who coined the fanciful name "Jayhawkers" became, six months later, the core of the Death Valley Jayhawkers, whose quest took them through uncharted lands and caused untold suffering before they could begin to toil for gold.

These 1849ers created their Jayhawker fellowship several years before the existence of the better-known Kansas Territory Free State Jayhawkers who fought proslavery militants near the Missouri border between 1854 and 1861. When John Colton, one of the '49er Jayhawkers, was asked why they called themselves by that name, he responded, "The Lord only knows. I suppose that it was about the most devilish name that could be fished up" (February 5, 1895).[1]

The California Gold Rush of 1849 was a time of hope, hardship, trepidation, and danger—death for some, but glory for others. The travels of emigrants from safe, secure home life into the wilds of the unconquered West read like a novel of adventure, mystery, and romance all rolled into one. They saw distant vistas, fields of wildflowers, and startling sunsets and shared singing, dancing, and new skills learned. But they also experienced dust, hailstorms, illness, injury, boredom, thirst, heartbreak, exhausting labor, and the death of comrades.

The journey from civilization to the goldfields was experienced by thousands, but only one particular group (fewer than one hundred men, women, and children) found themselves funneled in a narrow, obscure desert sink—Death Valley, the lowest place on the continent—and all but one of them (Captain Richard Culverwell) survived passage through that deepest valley in their travails when they were lost in the desert wastes of Nevada and California. The Jayhawkers were part of this group that has become known as the Death Valley '49ers. West of Death Valley, three more men died—from thirst, starvation, and broken spirits. The remainder experienced untold agonies. But from the horror of the desert their final deliverance was ecstasy.

Before the '49ers entered that lowest place on the continent, they had already traveled two thousand miles by wagon. Some claimed little occurred worth mentioning while crossing the plains, but that was because the tragedies and suffering of their final three hundred miles stood in stark contrast to all that went before. Thus, to illuminate their trek west from Illinois to Salt Lake City, I use the perceptive words of a fellow wagon-train mate, William B. Lorton.[2]

After the decision was made at Salt Lake City to create a wagon train to break a road southward to Los Angeles (the Southern Route), thence north to the goldfields to circumvent the snowy Sierra Nevada, various groups, families, and single men started gathering along Hobble Creek, about sixty trail miles south of Salt Lake City at today's Springville, Utah. This unwieldy train of ultimately 107 wagons became the San Joaquin Company, named for their final destination. Over time the name was logically corrupted to the Sand Walking Company (Manly 1894, 108).

Two hundred miles down the Southern Route, the Jayhawkers still exhibited their sense of fun and energy. When the wagons were camped atop the Black Mountains south of today's Beaver, Utah, Adonijah Welch reported he was "struck with the glare of a hundred fires and the motley groups busily engaged around them. Near one however the Jay Hawkers are dancing in grotesque attitude to the twang of an old fiddle" (October 26, 1849, in Ressler 1964, 268).

In southwestern Utah, at a place the emigrants called Mount Misery (some called it Poverty Point), these Death Valley '49ers (with about 26 wagons) found a way around a labyrinth of canyons that had stalled the San Joaquin Company and caused most of the wagons to turn back to the Old Spanish Trail. The Death Valley '49ers continued westward in their search for Walker Pass, an Indian trade and horse-thief trail through the southern Sierra Nevada to the San Joaquin Valley, a supposed shortcut to the California goldfields. These emigrants included four families, most of the Platte River Jayhawkers, the Bug Smashers (a loose collection of men mostly from the Bug Smasher Division of the San Joaquin wagon train), and various other individuals or messes of men who bearded the wilderness.[3] They wound their way through the mountains and valleys of Nevada, and by the time they funneled into Death Valley in eastern California, many had already abandoned their wagons or cut them into carts. They were reduced to short rations, and their cattle were in poor condition as the desert country plunged into winter. This was territory never before traveled by white men. All they knew for sure was the mighty Sierra Nevada blocked the way west where its southern reaches were penetrated by Walker Pass.

When the Jayhawkers (now expanded to about thirty men) entered Death Valley, they turned north to find a way over the western mountains, as did the Bug Smashers (about twenty-two men), the Brier family (five), and several independent men (about twelve).[4] Thus, I call this group the northern contingent of Death Valley '49ers. There was also a southern contingent (about twenty-eight men, women, and children) that turned south when they entered Death Valley.

The final dreadful phase of their long journey—the desert miles from western Nevada into Southern California—is the segment that deeply impressed the Jayhawkers for the rest of their lives. They suffered so much; they came so close to death by thirst and starvation that many could not think about it years later without weeping.

The Platte Valley Jayhawkers

No list has come to light delineating the men who were initiated into the Jayhawkers in the Platte River valley, but one can be assembled based on the Knox County Company roster plus lists created years later by John Colton (the youngest Jayhawker) and others who attended the Jayhawker reunions.[5]

Below are names of thirty men who *may* have been initiated into the Platte Valley Jayhawkers. Price, Montgomery, and McGowan did not continue west at Mount Misery in southwestern Utah, but twenty-seven of the Platte Valley Jayhawkers did become Death Valley Jayhawkers.

George Allen	John Hill Cole
Edward F. Bartholomew	John B. Colton
Bruin Byrum	Urban P. Davison
W. Carter	Edward Doty
Charles Clarke	Marshall G. Edgerton
Alonzo C. Clay	Sidney P. Edgerton

Harrison B. Frans

John Groscup

Captain Asa Haynes

Aaron Larkin

Edward McGowan[6]

Thomas McGrew

Charles B. Mecum

Lorenzo Dow Montgomery

John (or N. D.) Morse

Alexander Palmer

John Plummer

Robert C. Price

Luther A. Richards

William Robinson?

William B. Rude

Thomas Shannon

John L. West

Leander Woolsey

Included in the above list are some of the "elderly" men (over thirty-five years of age!) who became Death Valley Jayhawkers and whose names may be misplaced in this Platte Valley list. Asa Haynes (forty-five) was voted captain of the Knox County Company wagon train (and later its colonel), and it is unknown if he allowed the youngsters to initiate him into their merry band. John Plummer (forty-three), Marshall Edgerton (thirty-nine), and his cousin Sidney (thirty-eight) were also of an age where they may not have been interested in the high jinks. However, these older men were part of the Jayhawker Division of the San Joaquin Company that traveled the Southern Route from Salt Lake City to Mount Misery. Here these older men also turned west and became Death Valley Jayhawkers, possibly feeling responsible for the younger men.

Portraits of many of these men (and others who later became Death Valley Jayhawkers) are included. However, keep in mind that John Colton solicited them many years later when the Jayhawkers had become old men—no longer the carefree, naive young bucks of 1849. Only Alex Palmer's portrait had been taken before he left, and John Colton's two early ones were taken shortly after he arrived in California.

William Robinson is questionable as a member of the Platte Valley Jayhawkers. To date, the only information crediting him in that group is John Colton's wagon list of 1908 listing Robinson as traveling in wagon number eight from Knoxville (with John West and John Plummer, both Platte Valley Jayhawkers) (De Laney 1908, 101).[7] Thus, Robinson is included with a question mark. He died during the trek after crossing the Mojave Desert (Colton 1897). Robinson may have come into Salt Lake City in another train, as did Lorenzo Dow Stephens and Sheldon Young's mess, who also became Death Valley Jayhawkers. Ed Doty's son gives credence to this idea: "Along the trail wagons of other trains were passed and members of various groups became acquainted with each other. Among those that Doty thus learned to know were the brothers of his future wife, Mary Ann Robinson" (H. Doty 1938).[8]

Sources

A comprehensive book about the northern contingent of the Death Valley '49ers is now feasible because original copies of two daily diaries have become available—one by William Lorton, who traveled in the same wagon trains as the Jayhawkers from Illinois to southwestern Utah, and the other by Sheldon Young, who became a Death Valley Jayhawker. Lorton's journals, written while on his trek to California, provide a valuable, detailed record of incidences the Jayhawkers also experienced while crossing the plains. Young's diary is indispensable for its information of the trek after the Jayhawkers leave Mount Misery (in southwest Utah) and travel through the Nevada and California deserts. Although his diary is brief, it is the closest we have to a day-to-day account of that segment of their journey. Lorton's and Young's diaries provide an excellent scaffold on which to assemble the Jayhawker story from Illinois to Southern California. The discovery of Young's original diary is amazing, and that story is told in Appendix F.

I rely heavily on several works of William Lewis Manly, including "From Vermont to California," his serialized articles published between June 1887 and July 1890; his book *Death Valley in '49* (1894); his several essays collected in *The Jayhawker's Oath* (Woodward 1949); and his letters to the Jayhawker reunions. The reasons his work is reliable and invaluable are discussed in Appendix F.

An indispensable resource is the collection of original Jayhawker letters found in the "Jayhawkers of '49 Collection" at the Huntington Library in San Marino, California, originally compiled by Jayhawker John B. Colton. The collection includes more than eleven hundred letters, plus news clippings, scrapbooks, photographs, and maps written for or about the Death Valley Jayhawkers. Many of the letters were written by the Jayhawkers for their reunions, organized by Colton and held from 1872 to 1918. More about the reunions is available in Appendix C.

Lorenzo Dow Stephens's book, the diaries of Asa Haynes and Louis Nusbaumer, and the log of Adonijah Strong Welch, plus numerous contemporary letters, are also used to stitch together the routes these pioneers trod and to illuminate in their words the experiences and hardships they suffered on their trek to the California goldfields. A broader discussion of these sources is in Appendix F.

A word of caution about sources: The Jayhawkers' recollections morphed over the years, while newspaper articles varied in their accuracy. Jayhawker children and grandchildren inserted additional inaccuracies into the record, and researchers, who often had limited resources, did their best but sometimes could not help but skew the story. I have been exceptionally lucky to have so much primary material at hand to help present a captivating story as accurately as possible. There is something about Death Valley that encourages the commingling of truth and fantasy, and I have sought to separate the two throughout the text. However, some historical fantasies deserve more space, and they are addressed in Appendix D.

Editorial Procedures

Asa Haynes spelled his name *Haines* until the 1880s, when he changed it to *Haynes*. To eliminate confusion, I have standardized the spelling to *Haynes* (except in quotes).

The numerous quotations are handled in the following ways. For newspaper accounts, I silently corrected typographical mistakes. With handwritten letters, I adhered as closely as possible to the original spelling. I did not change the case of letters that began a sentence except at the beginning of quotes, and I added periods at the ends of some sentences and after some abbreviations and initials. When a word was repeated, I removed the repetition. A major exception to the above changes is Sheldon Young's diary. He wrote mostly in sentence fragments, so instead of adding periods, a raised dot is inserted at appropriate places.

With Lorton's original diaries, I worked with LeRoy Johnson's high-resolution digital photographs, with microfilm supplied by the Bancroft Library, and, most important, with the typescript made in the early 1960s by western history titan Dale L. Morgan. Morgan wrote out the "-ing" or "-ed" at the end of words where Lorton left a squiggle. Often Lorton's dates are off by a day. In those cases, the corrected date is added in brackets. I left Lorton's apostrophes as written (for example, "did'nt") rather than changing to today's usage. Parentheses in quotations are original. I used brackets to add letters, missing words, locations, and conjectural interpretations. A question mark in brackets indicates I think the word quoted is accurate, though the penmanship cannot be deciphered positively. Damage to a document that obliterated text is labeled [*tear*].

Citations in the text in parentheses correspond to the bibliographical references. In the bibliography all references from the Jayhawkers of '49 Collection are abbreviated as JCHL, followed by their specific collection number—for example, JA123. Maps are listed under "Map" in the bibliography followed by the date, title, and cartographer's name.

Acknowledgments and Credits

The most important person who contributed to this work is my husband, LeRoy Johnson. His dedication to research, experience in the field, work on the ground for forty years, and carefully organized personal library have been a large part of this book. When I refer to *we* in the text, it usually refers to LeRoy and me. I am also privileged to use LeRoy's excellent photographs to illustrate the text.

I extend sincerest thanks to Genne Nelson, Death Valley National Park scientist, who deciphered and recorded hundreds of Jayhawker letters and John Colton's first two scrapbooks—a prodigious amount of accurate work. She permitted me to use that material for reference. I also greatly appreciate David Duerr, who provided his scans of Sheldon Young's original diary and for his quintessential story about discovering Young's diary. My sincere thanks also to Jennifer L. Martinez and Brooke M. Black, who organized the Jayhawkers of '49 Collection in 2000, and Peter J. Blodgett, the H. Russell Smith Foundation Curator of Western Historical Manuscripts at the Huntington Library. All the Jayhawker quotes come from this collection, and the Jayhawker portraits come from the Huntington's Rare Book Department. I thank the Bancroft Library for access and permission to use William B. Lorton's California journals and letters and, for comparison purposes, Dale Morgan's typescript of the journals. My gratitude also extends to other libraries and keepers of official records dedicated to preserving and making available material from our western American past. I had help from professional genealogists Jo Linn Harline, Robert Hoshide, and Diann Biltz as well as Jayhawker descendants Arlene L. Doty, E. B. MacBrair-Koller, and others. My gratitude goes also to the interested and supportive persons, including Elden Hafen, Tom Sutak, Richard Bush, and so many more whose names have slipped into the past; all have helped with information and experience. There are those who have given me sage advice and whose appreciation of carefully researched books on western history have lent indirect support and encouragement—Bob Clark, Will Bagley, and supporters of the conferences on Death Valley history and prehistory that included many Death Valley '49ers, Inc. and Death Valley Natural History Association members. Discerning readers and friends have lent their time and expertise to improve this book. My sincerest appreciation to our two sons, Eric and Mark, who spent numerous childhood holidays tracing various combinations of the Death Valley emigrant trails through desert canyons and dry washes. Finally, I extend my appreciation to those dedicated researchers who have preceded me.

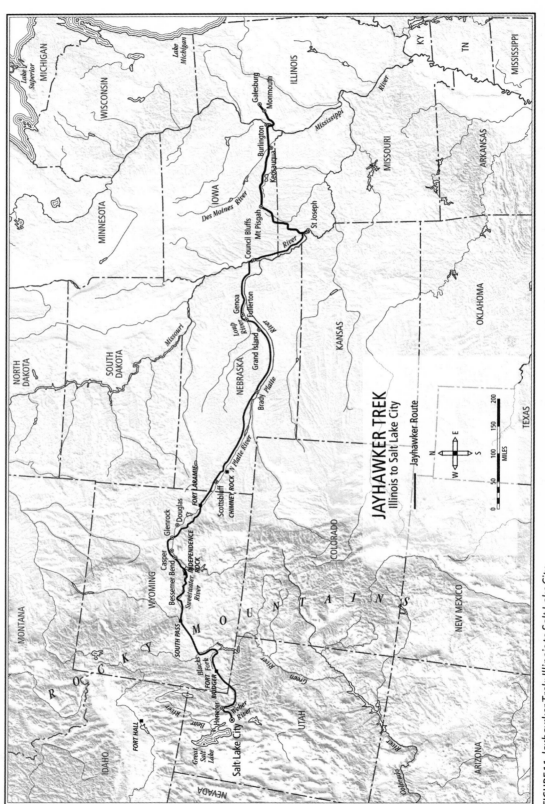

FIGURE 1.1. *Jayhawker Trek: Illinois to Salt Lake City*

CHAPTER 1

JAYHAWKER BEGINNING:
ILLINOIS TO COUNCIL BLUFFS

THE GOLD SEEKERS WERE "well equipped with oxen, wagons and provisions, high hopes and buoyant spirits.... Remember how jovially these ardent explorers used to sing the familiar refrain 'Then Ho! boys, Ho! and to California go; There's plenty of gold in the world untold, On the banks of the Sacramento'" (*Daily Iowa Capital* 1896).[1]

On April 1 (the day to send someone on a "fool's errand"), four fat oxen stood patiently before a prairie wagon in front of the Colton Mercantile Store in the western Illinois town of Galesburg while Sidney Edgerton, John Cole, and seventeen-year-old John Colton loaded supplies into the wagon that would become their home for the arduous cross-country trip to the California goldfields. John Colton remembered their sendoff years later:

> Crowds of friends gathered to wish the gold-seekers Godspeed. All wore smiles, and bits of humor of the time mingled in the conversation, but behind it all there was a full sense of the seriousness of the undertaking....
>
> Amidst the smiles and farewells of friends the church bell began to toll. As the party turned and looked in that direction they saw the village wag in the belfry...and as the men drove away, thought it fit and proper that he should toll the bell, as many never expected to see the gold-seekers again. They expected them to lose their scalps on the plains long before they should reach the region of gold. But the adventurers did not view the omen so seriously. One of them looked back and smiled. "It is April first—All-Fool's Day!" (De Laney 1908, 100–101)

The lure was *gold*—the siren's call—tempting men from across the nation to rush to California for easy riches, adventure, a better life—a fool's errand. The *New York Herald* spread the news to the whole world in the fall of 1848: "It is beyond all question that gold, in immense quantities, is being found daily in this [newly acquired] part of our territory.... Every vessel which anchors in the

FIGURE 1.2. Typical emigrant wagon, often a modified farm wagon, had a box usually
10 by 4 feet with canvas cover, often oiled and stretched over hickory stays. The
front wheels were about 3.5 feet in diameter and the rear wheels about 4 feet.

neighborhood of California, is immediately deserted by her crew... affected with
the mania... of gathering the rich material."

The *Oquawka Spectator*, read in the western Illinois towns of Knoxville,
Galesburg, and Monmouth, trumpeted on November 28, 1848: "One man dug
twelve thousand dollars in six days and three others obtained in one day thirty-
six pounds of the pure metal. Two months ago, these stories would have been
looked upon as ludicrous, but they are now common occurrences."[2]

In December 1848, an official report and a small tea caddy containing gold
dust and nuggets arrived in Washington, DC, from California.[3] James Marshall
had discovered seemingly limitless and easily obtainable gold on January 24,
1848, in the Sierra Nevada foothills of Mexico's Alta California. Nine days later,
the whole of California became part of the United States with the signing of the
Treaty of Guadalupe Hildago—part of the spoils of the Mexican-American War.

The world went wild when the golden contents of the tea caddy were put
on display. "The value of the gold in California must be greater than has been
hitherto discovered in the old or new continent," extolled the *Washington, DC,
Daily Union* on December 8, 1848 (Bieber 1937, 126). From this little tea caddy

erupted the largest voluntary mass movement of people the United States had ever witnessed—the California Gold Rush.

The torrent of humanity started west when men from every hamlet, from Maine to Iowa, reacted to the news of easy wealth. The frenzy was fueled by newspapers such as the *Oquawka Spectator* that reported on March 14, 1849: "Nearly every steamer passing...seems to have on board more or less persons on their way to the gold regions. The fever is decidedly on the increase. Almost every town in the country will be represented in California.... Lawyers, doctors, editors, preachers, painters, mechanics of all kinds and farmers alike are pondering over the matter, and some of them packing up their duds for a start." Earlier, on February 7, the *Spectator* reported, "The great attraction, however, of late weeks has been California. This is the El Dorado of avaricious expectation. The news comes that $40,000 are taken out daily from the gold washings there. Such tidings turn some heads hitherto reported sound."

A young man visiting relatives near Monmouth, Illinois, read these stories and was infected by the hysteria. William Birdsall Lorton decided "on going to California, every body talking of going, gold feever rages through out the Country. Commence buying thing[s] such as: spades, shovels, & tinware" (January 10, 1849). He responded to the following announcement in the *Oquawka Spectator* that all "who intend going to California in the spring by the overland route, are requested to meet" to organize the expedition (January 31, 1849).

Years later Jayhawker Charles Mecum (1878) wrote to his comrades about their departure: "A large party of young robust men,—well equipped, bade fare-

FIGURE 1.3. Galesburg, Illinois, in the 1870s showing the Colton store.

well to home and friends on that eventful 5th of April, 29 years ago:—*buoyant* with high hopes of the future: our goal the land of gold. Of our slow and tedious march for weeks through mud and mire—of our dances, our shooting matches, you all know."[4]

Jayhawker Urban A. Davison wrote in 1899: "Everything is fresh in my mind that transpired from the time we first left Galesburg and our parints & friends came out and campt with us the first night.... Our Hearts was light and gay and full of fond Hop[e]s. little did we think [of the trials] that we would Have to under go before we reached the land of Gold as we never Had campt out and never new what it was to be in need of or deprived of any thing that we desired." Sheldon Young was more brief in his diary: "Started on our trip for the gold regions · had pleasant weather but muddy roads" (March 18, 1849).

Dr. Ormsby, who traveled in the same train on the Southern Route south from Salt Lake City as did the Platte Valley Jayhawkers, summed up the mass movement:

> This emigration may be regarded as one of the greatest anomalies in the history of man. In enterprise and daring, it surpasses the crusades, Alexander's expedition to India, or Napoleon's to Russia. These originated with Potentates and religious enthusiasts—were supported by the combined wealth of nations, and urged forward by the power of Kings, Emperors, Pontiffs.... This is the spontaneous exhibition of the enterprise and daring of the American character.... Acting on obedience to no political power—urged on by no religious zeal, he is actuated solely by his love of adventure, and his love of gold. (in Bidlack 1960, 37)

Over countless centuries, famine, flood, and war have coerced mass migrations of people, but this western migration was different: *it was caused by personal choice*.

"The compulsion to see what lies beyond that far ridge...is a defining part of human identity and success.... Perhaps its foundation lies within our genome," wrote David Dobbs (2013, 50). He continued, saying a variant of the gene DRD4-7R is tied to "people more likely to take risks; explore new places...and generally embrace movement, change, and adventure and is found in about 20 percent of the human population." Maybe our Jayhawkers were some of those spurred by "curiosity and restlessness," to seek "both movement and novelty."

The environment in North American culture was ready for this burst of movement, change, and adventure. The nation had just acquired over 545 million acres of land after the war with Mexico, explorers had mapped routes to the West, mountain men had probed this wilderness, and early emigrants had successfully settled in distant Oregon. A fusion of nature (genes) and nurture (culture) was ignited by a little tea caddy full of gold.

The Gathering

On March 14, 1949, six weeks after the *Oquawka Spectator* published its notice "to take measures for organizing the expedition," it announced the starting date:

HO! FOR CALIFORNIA

All persons belonging to Capt. Findley's company are hereby notified to be in readiness to start for California by the 26th of this month. It is thought best to travel in companies of six wagons through to St. Josephs, Mo. *By order of the Captain.*

John Colton's father, Chauncey, had intended to have his teenage son travel with Findley's company. A letter written by one of the train managers said, "I made the necessary arrangement for his [John's] team and sent word to you to that effect at the time.... Yours was the only team I had any positive knowledge of as going from Galesburg, and from that fact I could make no arrangement for any other. Yours therefore will be the only team expected from your place" (Brockelbank 1849). At the end of this letter in John Colton's hand is this note: "This was the Company I intended to start with but we formed with Knoxville & others." At least eighteen men who became part of the Knox County wagon train had planned to go with Findley and later tried to catch up to him or convince him to change his plans and merge with the Knox County train. A Findley train member wrote, "The Knoxville teams have not joined us, nor will they. After we crossed the [Mississippi] river at Burlington a delegation from them overtook us, for the purpose of persuading us to abandon the idea of going to St. Jo., stating that it was their design to make a direct move to the Council Bluffs, and wished us to do the same" (W. [Newton Wood] May 16, 1849).

The company became known as "Findley's lightning train from Oquawka" because of the rapid pace it set. "It is said to be the quickest trip ever made by ox teams" to Fort Laramie, boasted H. M. Seymour. "On the 12th day of August [the Findley train] arrived at the Gold Diggings in California," six months *earlier* than the Knox County boys who became the Death Valley Jayhawkers (W. [Newton Wood] June 5, 1849; Seymour June 5, 1849; Findley August 16, 1849).

John Colton wrote his father when he was about forty miles west of Burlington, Iowa, on April 15, 1849:

The Ferry Boat came over [the Mississippi River] with the Knoxville Teams which we have joined and are going through with as Capt Findly would not go to Council Bluffs & St Joseph was chuck full. we have 18 teams and we think that we are strong enough to go through when we started from Burlington. each waggon put in $1. per yoke to bear expenses to Council

Bluffs but our Team is going to seperate on account of feed being so high.

We think that we can get along cheaper to go 2 Teams together.

As it turned out, these teams did not separate, but his statement makes clear John's wagon and the Alonzo Clay mess (with whom John rendezvoused at the Clay farm) were not originally part of the Knoxville group, nor were they contemplating staying with that group. Thus, the claim made in the *Death Valley Magazine* in 1908 that young John Colton had organized a train of twelve wagons is inaccurate (De Laney 1908, 101).

In 1888 John described his start for the goldfields to a reporter:

I was a lad of 16 [17] in 1849 when the gold fever broke out, . . . and it was not long before I had the disease in its most malignant form. I lived with my parents in Galesburg, Ill., and what was my delight one fine spring day to hear that a party had been formed to make the trip to the Golden Gate. I soon managed to get my name enrolled as one of the company, and a sturdier set of young fellows never packed up their little plunder and set out in search of their fortunes.

As afternoon waned on April 1, 1849, John Colton's mess traveled three miles from Galesburg to the Clay farm, where they joined Alonzo Cardell Clay (twenty-one), "Deacon" Luther A. Richards (thirty), Charles B. Mecum (twenty-six), and Marshall G. Edgerton (thirty-eight), who were the men in wagon number two, as Colton called it years later. All these men became Death Valley Jayhawkers. "At Clay's they met their first disappointment and were given an idea of what the traveler in those days might expect in the way of delays. They had scarcely pitched camp at Clay's when a blinding rainstorm set in. For five days and nights it continued with unabated fury. The men were only three miles from home, but they would not return. They had said they would return with the gold dust, and would not now show their faces even for a brief spell without the yellow metal" (De Laney 1908, 101).

Galesburg was a well-established town by 1849, having been settled in 1836 with a dream of George Washington Gale to bring a civilized, educated eastern village to the sod prairie of the "Far West" "through a system of mental, moral and physical education." The town has always been known for its educational institutions; it also became an active depot on the Underground Railroad.

Galesburg, with about four hundred inhabitants, impressed William Lorton, an outgoing and inquisitive twenty-year-old New York City house painter who was visiting cousins near Monmouth, Illinois, in late 1848: "pritty town, every thing looked clean. . . . Fine church, beautiful finish. Academy, college & schools at Galesburg. Fine society at this place. Academy brick, paint cream color. Cupalo on to[p]" (October 7, November 19, 1848). Several members of what became the Knox County Company wagon train were from the Knoxville area, the heart of flourishing farm country. Although Knoxville, three miles southeast of Galesburg, was the oldest town in the county and the county seat, it was uninspiring to Lorton.[5] Only the county courthouse, with its "steeple on top, 4 Ionic collums, brick & morter," was worthy of his attention (October 7, 1848). The wealth was in the surrounding farm and grazing land. For instance, Asa Haynes, later elected captain of the Knox County Company wagon train, had a thousand acres of land, had built a brickyard and sawmill by 1840, and had a twelve-room, two-story brick house with the largest frame barn in the county. He married Mary Gaddis in 1830, and she bore eight children (Ellenbecker 1993, 5).

The start of this momentous trek in search of gold was not propitious—mud, rain, and swollen streams caused slow progress. Lorton summed up the weather—"bad going"—as teams collected near Monmouth to form a train to the goldfields. From April 3 through April 7, he said, "Rainey & mud. . . . Rains hard. Cold & chilly. . . . Rains pitch forks [with] thunder & lightning. . . . Knoxville teams arrived last night. . . . More mud to start with than I ever see before. As the teams pass through, the mud roled like the waves in the sea." And the Knox County boys were not yet out of Illinois!

To add to the gloomy mood were reports of John Frémont's disastrous fourth expedition, in the winter of 1848–49. Ten of the thirty-three members died from starvation and exposure in the San Juan Mountains in Colorado. A few of the expedition members resorted to cannibalism in order to survive, which was widely reported in the nation's newspapers. The lurid stories of cannibalism suffered by both the Donner Party the winter before and the Frémont disaster made argonauts leery of the required crossing of the Sierra Nevada, the mighty barrier on California's eastern border, especially late in the season. It caused a mood of impatience to win the race with the menacing onslaught of winter. The thought running through their minds was almost visual—Gold! Gold! Gold! Hurry! Hurry! Hurry!

On Friday, April 6, 1849, Asa Haynes wrote, "Traveled 16 m. whare we fell in with the rest of the company in Monmouth." Here "the company was to form some sort of an organization, hire a pilot, and choose a Captain, but for some reason this was not done and on the morning of the 7th we left Monmouth, then a small village, and started westward." They had "33 men & 12 waggons," soon augmented to eighteen wagons (Haynes, April 6–7, 1849; Wiley 1937, 1).[6]

Haynes was chosen to lead the wagon train and was known henceforth as captain (Wheat 1939a, 105). At age forty-five, he was older than most in the train and was a prosperous businessman. Asa leased his land and put his family in the care of his nephew before traveling west with his close neighbors Bruin Byram and Edward Doty (Haynes n.d., 3).[7]

William Lorton gives an idea of the contents of the Knox boys' wagons: "Our team consists of 4 yoke oxen & 1 [yoke] cows (non milch)...about 2500 [pounds] on board the wagon, 1000 flour & meel (kill dryed), 350 Bacon, Teas, sugars, Coffee, Lemon syrup, medicines, dryed fruit, nearly all the provisions in bags, [and] bed quilts" (April 2, 1849). Numerous quotes from Lorton's daily California trail journals help illustrate the trek from western Illinois to Utah.

Lorton's mess consisted of himself (age twenty); Edward Kellogg (fifty); Edward's cousin Frederick Saxton Kellogg (thirty-nine); Edward's son William (about Lorton's age), who had just completed a year at the Galesburg Seminary; and Henry J. Ward (also about Lorton's age), who was Edward Kellogg's son-in-law.[8] Ward worked at Lorton's cousin's mill, and he and Lorton had both signed up in February to go with Findley's train, but delays caused them to miss it.

When the moment for parting finally came, it was more difficult than expected: "How the tears did flow at the Hand shaking and fare well Biding in the Morning when we Bid our Friends Adue," recalled Urban Davison in 1899. "Hard parting, good deal of crying. I stood it without a tear, hard hearted, I suppose" (Lorton April 2).

Just before he left home, Edward Doty, then twenty-eight, wrote his "last will and testament to be in full force and effect, providing I never return from California" (April 2, 1849). Doty was probably one of many who worried about surviving a venture where being accidentally shot, run over by a wagon, attacked by Indians, drowned in a river, or struck down by cholera were potential ways of dying on the trail.

Others were having second thoughts. John Colton wrote from Kanesville on May 22, 1849, that "Sid [Edgerton] wanted to go home so bad that I finally told him that he might sell & I would buy a share in Marshalls [Edgerton] waggon. as we got [out] of town [St. Joseph] though, he began to feel better."

William Lorton wrote about the departure: "April 6. Knoxville teams arrived last night. They camped south of Monmouth. We visited their tents, all struck [set up], camp fires & stoves all burning. 1 marque tent, table set, stools &c. With pistols in belts, knives &c. & dresses, looked like a scene in theatre. April 7. Come up to K[noxville] teams at Cecedar Turnbulls. Word comes that Knox C comp had struck [taken down] tents & started.... Have dancing in eve'g, violin & singing." From Monmouth the route was west to Oquawka on the Mississippi River, then south about seven miles to Shokokon. Asa Haynes's diary, April 9: "We traveled

20 [m]. encamped 2 miles east of Shockcon."[9] April 10 they "ferried the mississippi up to Burlington. encamped one mile west of town."

The *Oquawka Spectator* mentioned some of the teams in the large contingent
that stretched from Monmouth to Galesburg (names with an asterisk became
Death Valley Jayhawkers):

> On Monday last [April 9], there were fourteen wagons passed through this
> place from Knox and Warren counties [Illinois]—having camped a few
> miles from the town on Saturday last. This company have commenced
> their long journey by "keeping the Sabbath day."—May prosperity attend
> them.
>
> *Henderson.*—Nelson D. Mor[s]e,* Orrin [Oren] Clark,* James E. Hale—One
>   Wagon.[10]
>
> *Knoxville.*—Alexander Ewing, Jno. Ewing, L. Dow Montgomery, Robt.
>   Rice [Price?], A. M. [Luther A.?] Richards,* Edward McGowan—Three
>   Wagons.
>
> Josephus [Cephas] Arms, E. [Norman] Taylor—One Wagon.
>
> J[ohn] L. West,* J[ohn] W. Plummer,* A[sa] Haynes,* Thomas McGrew,*
>   George Allen*—One Wagon.
>
> Robt. Kimball [Kimble], D. C. Norton, Nathaniel Hulbert [Hurlbut], Wilson
>   Temple, Leander Woolsey*—Two Wagons.
>
> Alex. Palmer,* Aaron Larkin,* Thomas Shannon*—One Wagon. (April 11,
>   1849)

A list of departees from Galesburg and neighboring towns was also remembered late in life by John Colton. His list of wagons was first published in 1908
in the *Death Valley Magazine*, from Rhyolite, Nevada. It included several men
who turned back to the Old Spanish Trail in southwestern Utah and thus did not
become Death Valley Jayhawkers. Colton's list adds to, rearranges, and duplicates
some of the men in the *Oquawka* list above.

> In wagon No. one, was John B. Colton, Sidney P. Edgerton, John Cole;
>
> No. two, Alonzo C. Clay, "Deacon" Luther A. Richards, Charles B. Mecum,
>   Marshall G. Edgerton;
>
> No. three, Harrison Frans, John Groscup, [from] Henderson Grove;
>
> No. four, Charles Clarke, Urban P. Davidson [Davison], John Morse of
>   Henderson;
>
> No. five, Capt. Asa Haines [Haynes], Thomas McGrew, George Allen, near
>   Knoxville;
>
> No. six, Alexander Palmer, Thomas Shannon, Aaron Larkin, Knoxville;

No. seven, Edwin Doty, William B. Rude, Bruin Byram, near Knoxville;

No. eight, John W. Plummer, John L. West, William Robinson, Knoxville;

No. nine, Edward F. Barthlemow [Bartholomew], father and brother, of
   Farmington;[11]

Nos. ten and eleven, Alexander Ewing and son John C. Ewing, owners
   of two teams, and three other men they were taking through to the
   mines;

No. 12, Deacon Cephus [Cephas] Arms, Robert Price and Norman Taylor,
   Knoxville. (De Laney 1908, 101)

The *Oquawka Spectator* was probably more accurate for those who started
together, but over time, and certainly after Salt Lake City, wagon personnel
changed. Fifty years of memories and many Jayhawker reunions added still more
layers of change before Colton came up with his list above. Lorton said there
were eighteen teams when the train was in Iowa (April 13, 1849).

The Trail

While camped a mile or so west of Burlington, Iowa, the Knox County group
decided to ship their heavy goods down the Mississippi River and up the me-
andering Missouri to Kanesville, Iowa—a distance of nearly 800 miles—to ease
the weight the oxen had to pull for 300 miles across Iowa from Burlington to
Kanesville. They sent two men with the baggage, and another two men were sent
ahead to Kanesville, also called Council Bluffs (Lorton, April 11).[12]

The Knox County train had planned to travel west on the early government
roads in southern Iowa to Keosauqua, then follow the Mormon Pioneer Road to
Pisgah (today's Mount Pisgah) and on to Kanesville. This road strung together the
"Camps of Israel," which were Mormon rest stops. But near Pisgah the Knox County
train began a 130-mile detour southward to St. Joseph, Missouri, to retrieve the baggage they sent by boat, which had become stuck on a sandbar below St. Joseph.[13]

The company expected to take the Council Bluffs Road on the north side of
the Platte River after crossing the Missouri River in the Kanesville–Council Bluffs
area. Merrill Mattes said about the road:

Historians today are in the habit of referring to the trail north of the Platte
as the Mormon Trail, since the Mormons, or Latter Day Saints [LDS], primarily used that route.... This trail was actually blazed by fur traders in
the 1830s. Thousands of non-Mormons, or "Gentiles," used this north
route during the California and Montana gold rushes, and they probably outnumbered the Mormons in 1849.... Thus, in contemporary usage

there really was no distinct Mormon Trail from Council Bluffs to Fort Laramie. It was, instead, the Council Bluffs Road. (1969, 7)

A few quotes from William Lorton's journal give a flavor of their travels through Iowa:

[April 19, 1849] One place we decend, we unhitch 4 yoke & lock both wheels on account of the steepness of the place....[14] To hear the noise of the many teemsters bawling to their cattle, the lashing of whips & the greating of the tires over the stones & grit, it presents a scene beyond description.

[April 20] 1 man got killed in our rear by his comrades rifle. He was in the brush & the first thing he knew a ball passed through him. I have just rec'd news of another man ahead being killed in a similar manner.... Nearly all the comp. are out gunning. Bill [Kellogg] brings in 10 gray & fox squrrels which we have for supper.... Some of the men cuts down a bee tree & collect lots of honey from it.... Great fears entertained of the cholera in Keosauqua.

[April 21] Had a concert in our tent in the eve'g, & to help the base [bass], roleing thunder joins the chorus, while clashing through mid air. The rain is, meanwhile, decending in large drops.

[April 22] If the weather continues warm we may expect grass in a few days, & then we can persue our jorney.

[April 24] I was taken sick to day, & after taking a remedy went to bed. Here we had scarcely any water & what we had the cattle would not drink. We drank it.

[April 25] Bot some maple sugar of a woman. Along this road there is considerable travelling, & a no. of log cabins going up of considerable size, showing the march of improvement westward.... I turned in tireder than I had ever been in my life before.

[April 27] Killed 5 rattle snakes.... Came within 1 [ace] of setting our tent on fire by the leaves.... To day we made a 20 mile drive.

[April 29] The wind blows a hurricane, some rain & a little sun. Our lost men return—some ponies lost but get them.

[April 30] We have a wind storm, a pampero, a torpedo, a whirlwind, any thing you are a mind to call it. When you have to shovel the dirt out of your eyes, a day when the d[evil] is in the cattle, when the roaring whirlwind & the flakes of sand cuts your eyes nearly out, when you have to pilot & cant see, when every boddy is cross & ready to fight, imagine this & multiply it by 20 & you may get a faint idea of this day.... Make for Pisga 15 miles dist.[15]

Based on clues from Lorton's river names and the map of Iowa by Henn, Williams & Company, published in 1851, the Knox County group took the road due south from Pisgah along the East Fork of the Grand River on their detour to St. Joseph to retrieve their stranded supplies. At some point, possibly after entering Missouri, they swung westward to the Grand River (Lorton's west fork of the Grand) and soon headed west again to the little Platte River, possibly in the Parnell, Missouri, area. Lorton indicated, "We finally haul up on the [little] Plat[t]e River. Here there is quite a settlement, 3 log cabins & land under cultivation" (May 5).[16] They traveled down the little Platte, then crossed west to One Hundred and Two River not far north of Savannah, Missouri.

From Council Bluffs John Colton wrote about their problems with the baggage at St. Joseph:

> The reason of our not reaching this place [Council Bluffs] any sooner was that we sent Price around with our stuff, and as it was low water he was afraid it would not get here, so he stoped it at St Joseph & sent out an express, and we went in there about 120 miles out of our way. we got there & found our flour, but our bacon, picks, axes, &c Together with some of Clay's things were among the missing. Father wrote me by Price that he had insured it but that did no good. the steamer Brat had been there & left so we had to contend ourselves with paying 13.50 for freight & buying new Bacon. (May 22)[17]

On May 9 the train turned north from St. Joseph and traveled up the Missouri River bottom and bluffs to Kanesville. After all the mud and rain and the need to buy corn for the cattle, the blooming spring—green grass and May flowers—was a blessing. Lorton waxed poetic about "the rose just bordering the wild lilac in full bloom, the crab apple in blossom, & the wild plum bursting from the shoot...the sweet smell of the early flowers & freshness of the new green leaves above the grassey carpet, the cooing of the ringneck & turtle dove, the chirps of the robbin & molking bird" (May 11).

The Knox County Company was formally created on May 12, a month after they left Illinois and almost three weeks before arriving at Kanesville. The train halted to "form a more perfect organization" on a pleasant bottom near the Missouri River north of the Nodaway River while still in Missouri. Lorton announced with little fanfare: "12 o'clock have meeting of comp.... Pass constitution and by-laws" (May 12; see also Appendix A). On May 16 he reported, "All sign the constitution." They passed through Linden, Missouri, where "the residents are Mormons, generaly, & is a pritty frontier town.... Cross the Nishnabottona on flat boat. Take 1 yoke & wagon at a time. Ferriage 65 cts.... Lot of our men go spearing fish."

Near Kanesville, Iowa, the Knox County Company camped on the bottomland "west of the town, & 4 miles from the river on the edge of a creek" (Lorton, May 20). Men went into town to purchase last-minute supplies, have a meal, or even to take lodging somewhere other than a tent or wagon.

John Colton wrote friends via his mother, Emily, just before his company ferried the Missouri River on May 22, 1849: "I have paid Mr. Hyde [editor of the *Frontier Guardian*] for several copies of his paper which I send you one of others to Mass [Massachusetts]. you will find our names & constitution.... I was on guard ½ a night last night, so of course I do not feel very keen.... there has been two or three nights that we have had no wood.... we have had milk every noon while some of the rest have had to drink water. they have nearly all got cows now."

Colton wrote a more detailed letter on the same day (via James H. Notewan), giving an overview of their route through Iowa:

> As this is the jumping off place from civilization into eternal unciviliza-
> tion I employ this last opportunity of writing to you. it may be for some
> time, and although you would naturaly think that it being the last place
> inhabited by civilized people I should have feelings akin to Sorrow but
> this is not the case. I think I have seen quite enough of the Inhabitants
> of Iowa especialy, and I am itching for morning to come when we shall
> cross the river. Since I wrote to you at Washington [in southeastern Iowa],
> we have traveled over a great extent of wild country. about a week after,
> we had grass so that with what corn we had we kep[t] our oxen in pretty
> good spirits. they are now in better order than when we left home. They
> fattened up right away as soon as they come to have plenty of grass. we
> have not had out a guard yet but there is considerable many Indians this
> side [of the river]. tomorrow we cross & then we begin. we have a regular
> organized company consisting of 36 waggons. Asa Haynes of Knoxville
> is elected Captain. the Canton & part of the Roc[k] Island Company have
> united with us. we are going through under a military regulations....
>
> Iowa is full of nothing but Mormons and all going to the Salt Lake. they
> are perfect thieves. you have to look out for them or they will clean the
> cash out of you in a hurry—we had a first rate road while we was going on
> the Bluff road but when we turned on to the St Joseph road it was very hard
> for our cattles feet. There was some cholera in St. Joseph & there is [some]
> here.... We was offered 80 Dollars a yoke for our cattle [at St. Joseph] and
> could have sold our waggon &c for cost.... there was about 1500 waggons
> to cross when we was there & that looked rather gloomy. 2 Ferries were at
> work constantly. Ray left about 2 weeks before we got there. we stopped 1

week at Keasauqua to recruit our cattle, so they got the start of us.[18] this
is the last place a man will have the privelige of backing out as he can not
come back after leaving here 100 miles on account of Indians.

The Mormons needed money to buy supplies for their own travels to their new
Zion—Salt Lake City. Most of them were very poor and did all they could to make
enough money to leave the states.

At Kanesville, on May 20, Lorton noted the Knox County Company "had several applications to join ou[r] comp[any]." Their names were added to the roster
published in the *Kanesville Frontier Guardian* on May 30, and the new additions
doubled the size of the original company (see Appendix A).[19] Although the largest percentage of signatories came from Knox Country, the company included
emigrants from eight other western Illinois counties.

Colton neglected to mention the second largest subgroup, the Fayette Rovers
from southern Michigan, with sixteen men. They probably had four wagons to
augment the train for the ultimate total of forty-two wagons mentioned later.
The Rovers signed the Knox County Company roster shortly before the company
began to cross the Missouri River on May 22. Their captain, Henry Baxter, later became colonel of the San Joaquin Company on the route south from Salt Lake City.
Another Rover was Adonijah Strong Welch, whose letters and log of the Southern
Route in Utah provide some of the most accurate information about the trails.

On May 22 Lorton says, "In all we have 42 teams each averaging 4 yoke oxen."
If horses and milk cows are added, the train had more than 350 head of stock
that needed to be fed daily.

Twenty-four of the men who signed the constitution from the Knox County
area also traveled through Death Valley to reach the goldfields. They, with
Robinson and Richards (who are not on the roster), are among the men referred
to as the Death Valley Jayhawkers of '49.

Five of the company's elected officers became Death Valley '49ers—Asa
Haynes, Thomas Shannon, Edward Doty, Charles Mecum, and John L. West.
Captain Asa Haynes has already been introduced. Tom Shannon (twenty-five)
was a newcomer to Knox County, having moved there after his discharge from
the war with Mexico. He stayed in California. Ed Doty (twenty-nine) settled
in Knox County in the 1840s and, after panning for gold, stayed in California.
Charles Mecum (twenty-seven) came to Galesburg with two of his brothers in
1844 and returned after his California trip. John Lewis West (about twenty-nine)
remained in the West as a miner.

Some of the Death Valley Jayhawkers became prolific letter writers to the
Jayhawker reunions held many years after their trek to the goldfields. One was

Edward Franklin Bartholomew, age twenty-one in 1849. He was sixteen years younger than his brother, Luzerne, who traveled with him. The Bartholomew family had settled in Elmwood Township, midway between Peoria and Knoxville, where Luzerne was already married and had developed his own farm (Runner 1995, 1–2). Edward was closer in age to the young men who were initiated as Platte Valley Jayhawkers and remained with them when the brothers separated at Salt Lake City. Actor Clint Eastwood is one of Edward's descendants (McGilligan 1999, 14). Luzerne took the wagon and entered California via the Lassen Cutoff, arriving in Sacramento November 3, 1849. He became famous for the gigantic grizzly bear he captured and took East across the Isthmus of Panama. He toured the Atlantic states and principal cities of Europe with his bear, which outlived Luzerne and ended up in the New York City zoo.[20]

Another prolific writer was Lorenzo Dow Stephens, age twenty-one when he started for California. He was not a Platte Valley Jayhawker, nor did he travel in the Knox County Company, but he was from the Galesburg area and probably knew several of the "boys" from school. His family settled in Abingdon, eleven miles south of Galesburg and only three miles from Asa Haynes's farm, which was northeast of Abingdon, closer to Knoxville. Stephens's company left Illinois on March 28 and traveled across Iowa to Council Bluffs. About that portion of his trip he said:

> That spring happened to be a very wet one, and the roads were almost impassable.... bridges were washed away, and consequently much time had to be spent repairing and building new ones. On many of the larger streams we constructed rafts of logs.... Through Iowa we found many prairie sloughs, and they seemed bottomless. Here we had to cut sod and lay several thicknesses before we could pass over....
>
> When we reached the Missouri River at Council Bluff, we travelled down the river to Traders' Point, a distance of ten or twelve miles.

Near the Missouri River Stephens's company "remained for a week, waiting for the grass to get a good start, arranging for a larger expedition" before crossing the river (Stephens 1916, 8).

Thus, most of the Death Valley Jayhawkers were assembled in late May to cross from "the states" into "Indian country" on their trek to the golden streams of California.[21]

CHAPTER 2

MISSOURI RIVER TO SALT LAKE CITY

We were a goodly company to behold:
with our buckskin pants and bald heads.
—Charles B. Mecum, 1878

THREE FERRIES CROSSED the Missouri River in the Council Bluffs area: the Trader's Point ferry, ten or twelve miles south of Kanesville, serviced the Bellevue Indian mission on the west bank; the middle ferry, about nine miles southwest of Kanesville; and the Mormon (or upper) ferry that crossed about ten miles north of Kanesville to a point just south of the Mormons' Winter Quarters in Indian country. "For many years the principal crossing of the Mormons was not due west of Kanesville, but about ten miles north...sometimes referred to as Upper Ferry" (Mattes 1969, 124). The advantage of using the upper ferry was to avoid "the long and hazardous journey up steep bluffs that had always characterized the original" ferry site south of Kanesville (Bennett 1987, 48 [map], 49, 265n21).

About half the teams in the Knox County Company crossed the Missouri River at the upper ferry on May 23 (Arms 1849–50, also in Cumming 1985, 1). Others crossed their "cattle over [at] sunrise" (Lorton, May 24), and the remainder, held up by bad weather, crossed the following day.

Lorenzo Dow Stephens, who later became a Death Valley Jayhawker, crossed at Trader's Point:

> The ferry was a small scow and could carry but one empty wagon at a time. The scow was propelled by two oars, two men at an oar, and the current was very swift. Imagine the time it took to transport fifty wagons and the loads; we had difficulty getting the cattle to swim at first. We didn't realize that the sun shining on the water made much difference, so the first time the cattle swam round and round for two hours.... But next morning before sunrise we started a small boat with a couple of men

having a steer in tow, all the rest of the cattle followed without any trouble and made the opposite shore safely. (1916, 8)

As these young men were finally stepping into Indian country, Lorton took time to express feelings that were surely felt by the others who were "fixing there fire arms, loading their revolvers & rifles, sharpening their bowie knives & daggers, & preparing for service should it be required":

> The sensations produced by leaving a civelized country for a wild wilderness, by leaving a christian land for a barbarous country, one inhabited by the savage & unprincipaled red skin, launching out as we do upon hardships & troubles, dangers & perhaps death by a savage foe, a victim of the scalping knife, the tomehalk, or the arrow from a high strung bow, perhaps starvation, or wild beasts of prey may yet hold a caucus meeting over our tenements. we may have to encounter the eternal snows of the mountains & perish in the high hills of Cal. these are the sensations produced by entering the western wilderness. (May 24, 1849)

John Colton recalled thirty-nine years later, in 1888: "Until we reached Salt Lake our journey was uneventful. It was tedious, and some of us got homesick, but we had no adventures worth recording." However, William Lorton recorded a great deal he found of interest.

FIGURE 2.1. Missouri River north of Council Bluffs looking southwest.

The first Indian they saw in Indian country was a Pawnee: "He was well dressed & a fine looking personage. hundreds of their ponies died in the winter, the bones of which can be seen lying in the hollows" (Lorton, May 24). This was in contrast to the unkempt Indians they saw in the Kanesville area on the Iowa side of the Missouri. In the winter of 1848–49, many Indians suffered starvation in addition to losing their wealth—their horses—from the severe winter and wet spring that delayed the prairie grass, and the Sioux were still preying on the Pawnee, as they had been for years.

Dow Stephens's company had their first experience with Indians at the "first camp across the river."

Our camp fires were going nicely, supper was started, when we heard gun shots, volley after volley. In a few minutes from over the ridge came two to three hundred Pawnee Indians, riding at full run straight for our camp. It was a few minutes work for us to get our rifles in readiness, but the Indians put up a white flag, and they were allowed to enter camp. It seemed that a party of the Sioux tribe had given them battle, the two being at war, and the Pawnees had rushed to our camp expecting protection, but we ordered them off, telling them we wished no trouble with the Sioux as we had to travel their country, and wanted no enemies. We took the precaution to organize our body with regular military style with Colonels and Captains. (1916, 8)

FIGURE 2.2. Platte River. "We drink water out of the Nebraska [Platte River].... far better than slue water for the health, and less liable to produce diareah" (Lorton, June 11).

According to Charles Mecum, about four days after crossing the Missouri River, some of the young Galesburg-Knoxville men started their brotherhood rituals at the Elkhorn River camp, and the Jayhawkers were born. He recalled twenty-nine years later, in 1878:

> The first of the many little incidents I shall attempt to call to mind will commence after crossing the Horn river, where we were all barbered and took new names coresponding with those of the place we were in. How like boys we frolicked, and gave each other new names—J. B. with his head shaved—all but a scalp-lock—was our "Big Pawnee."[1] A. C. [Alonso Clay] with smooth head, and long beard was "Uncle Josh." Myself as the "Big Paddy"—and the little "Deacon" [Luther Richards] followed suit. We were a goodly company to behold: with our buckskin pants and bald heads. We needed no mirror to tell us we were handsome—*we just knew it.*

Lewis Manly got his description of the Jayhawker initiation rites years later from Jayhawkers Tom Shannon and Ed Doty when all three men were then living in the San Jose, California, area. He mentions the barbering occurred just after the Loup Fork crossing, on June 3, instead of after the Elkhorn crossing on May 27.

> These Illinois boys were young and full of mirth and fun which was continually overflowing. They seemed to think they were to be on a sort of every day picnic and bound to make life as merry and happy as it could be. One of the boys was Ed Doty who was a sort of model traveler in this line. A camp life suited him; he could drive an ox team, cook a meal of victuals, turn a pan of flap-jacks with a flop, and possessed many other frontier accomplishments. One day when Doty was engaged in the duty of cooking flap-jacks another frolicsome fellow came up and took off the cook's hat and commenced going through the motions of a barber giving his customer a vigorous shampoo, saying:—"*I am going to make a Jayhawker out of you, old boy.*" Now it happened at the election for captain in this division that Ed Doty was chosen captain and no sooner was the choice declared than the boys took the newly elected captain on their shoulders and carried him around the camp introducing him as the *King Bird of the Jayhawkers.* . . . however the boys felt proud of their title and the organization has been kept up to this day by the survivors. (1894, 321–22)[2]

In *The Jayhawkers' Oath,* Manly gave different details about the initiation ceremony:

> When they initiated a candidate, he was taken bodily on the shoulders

of four stout men after the evening meal was over, and carried around the camp fire on their shoulders, face upwards, and while the circuit was being made and the starting point reached, a song was being sung by those assembled in a circle, with a violin accompaniment by Mr. Isham. When a halt was called, the candidate's pants were shoved up on his leg and a good-sized bit of hair was pulled from his shin.[3]

Then the Captain of the Order asked the candidate if he would "solemnly promise to stand by his comrades through peace and war; when danger was in camp and under all circumstances. Remember this is your solemn oath; do you agree?" The answer was, "I do and will obey." The march was then continued until the second revolution was made. When the stopping place was reached, the candidate had to submit to the same painful operation on his leg, together with another provision of the "ritual," and if he did not squeal, the boys made the welkin ring with boisterous shouts and laughter. At the end of the third circle around, the finishing part was spoken, in slow and solemn tones, suited to a death warrant, and the candidate was declared a fit man to be a comrade. (Woodward 1949, 25)

The act of pulling hair from the shin may be why Charles Mecum referred to "that once great and glorious old band of Jay Hawkers and Shinpickers" (1872).

After crossing the Elkhorn River, the men met more Indians and then went hunting:

After walking 14 miles from the [Elk]Horn we strike the Platt. here we meet 2 Indians. they are friendly & we make them presents of pipes, tobacco, pork, bread, &c.... they want corn very bad, for they are planting over the river.... we bought a buffalo robe of the Indian for a bu. [bushel of] corn. The [Platte] river here is 100 rod[s] wide [550 yards].... killed 1 wolf, chased 1 elk, 1 antelope, & some see three more. double our guard to night. (Lorton, May 28–29)[4]

Stephens also saw wolves: "There were many different classes of wolves to be seen on the prairies: the common prairie wolf, the gray, the black and another, a large long-legged wolf,...the Buffalo Rangers, as these wolves were called. These wolves were ferocious, and a band of them would attack men, if hungry" (1916, 10).

Less than forty miles from the Missouri River, the emigrants began observing signs of distress along the trail. "We find several things thrown a way to lighten [wagons], such as irons, plows &c." And "dissension seams to have entered in the heart of the comp[any]. some are for proceeding & leaving those [latecomers] behind, & others to the contrary" (Lorton, June 1, 1849).

Lorton quoted William Findley from a meeting he attended February 7 about going to California. "As soon as a comp[any] gets beyond the settlements, no law to guide them but the law of the strongest, every person was as if he was let loose.... they complain within themselves, till at last they break out like a pent up volcano.... if the comp[any] was tired & f[at]igued, you might ask a man the most cival question & he would curse" (February 7, 1849). A wagon train was a community unto itself, and each developed a personality of cooperation and dissension depending on its membership and leadership.

Near today's Columbus, Nebraska, the trail left the Platte and followed the Loup River westward. On the last day of May, south of today's Genoa, Nebraska, they "ford Beaver River. Water comes up to [the wagon] box. The banks are very precipitous, but are dig[g]ed away. Great & exciting time in crossing." River and stream crossings were the most difficult part of the journey. Wagons tipped over, baggage and foodstuffs were damaged, a few men drowned (as did some oxen and horses), and it was cold, dirty work.

On June 2 Lorton described the Loup River crossing, "dreaded by all going this rout.... The streem is swift in the channel, & as the wagon moves along it grates & jars as if you were running over stones.... We squash a wagon wheel in our comp., lend [them] 1 of our wheels to cross the river with. Take off false tongues & make spokes, & the old wheel is made whole in the twinkling of a bed post."

FIGURE 2.3. Sketch by William Lorton. "CROSSING LOOP FORK, JUNE 2d 1849" west of today's Fullerton, Nebraska. Note the grave of Harrison Rowe on a hill in the foreground, killed by Indians three days before Lorton arrived. The serpentine route across the river is to find the firmest footing through the quicksand bottom. Banc MSS C–F 190 v 6, courtesy of the Bancroft Library.

After crossing the Loup River, they subdivided the train due to its increased size. "We elect 4 capt. & make our present capt. [the] Colonel, there being 40 wagons, 1 capt to every ten" (Lorton, June 3). This is when Edward Doty was chosen captain of the division afterward known as the Jayhawkers.

Monday, June 4, the company turned south toward the Platte River. "Travel over a heavy sandy road in the morning.... Carry brush 24 miles, the dist we travel today, to mend a bad slue or creek."[5]

This day was the beginning of a string of disastrous stampedes. In fact, over time, the Knox County Company was labeled the stampede train and shunned by other trains. The loss of oxen, loss of time hunting them, damage to wagons and equipment, and injury to men and stock were the cause of pain, suffering, anxiety, and the fear they would be caught in the deep snows of the Sierra Nevada in the far distant West. Lorton provides a vivid description of this first serious stampede on June 4:

> You turn & look & lo & behold the hind teams [come] running towards us like magic & as quick as lightning the spirit of "stampede" runs along the whole train.... I turn as quick as possible & find 4 teams within a few rods of me with scarcely room to escape. I pulled off my hat & succeeded in turning a team coming full split on our near side.... The whole train has now the epedemic, & like a mighty whirlwind they fly towards us. One team stumbles down & the next are thrown on top of that.... On they come like a mighty avalanch toward us, chains rattling, wheels grating, Wagons cracking, piles of oxen upon oxen, Men thrown down, cattle run over, & all in the utmost confusion.... We found as luck would have it that not much damage was done to the wagons, & only 1 cow seriously injured. This belonged to Capt [E. N.] Taylor, & his wife nearly cryed her eyes out.... Capt [Asa] Haynes was knocked down & hurt, one mans leg strained, & out of all the persons knocked down, none was very seriously injured.... Well has it been said the devil is in every thing upon the Platt.

Charles Mecum wrote to the 1878 Jayhawker reunion: "Those stampedes you have not forgotten. I can still in memory see a confused mass of cattle—wagons—women and children—one-horned cows &c going helter skelter—rattle-le thrash into the bottom of a ravine. I can still see the little Deacon [Richards], holding with *both* hands to the bows of the wagon—and bouncing up and down like a spile-driver, with such force as to go through our table."

As they traveled over the prairie, "the ground is litterally undermined by... creatures, such as wolf, prairie dogs, gophers, &c.... wade up to our knees in beds of flowers, flowers of all traits, shades, & hues, fragrant as the rose.... at last we come in sight of Grand Island.... Suddently large hail stones decend &...our cattle were driven by the hail some dist[ance] acrost the plains" (Lorton, June 6).

Four days later, they had another stampede, and this time Lorton's foot or ankle was badly sprained. Then in the early hours of June 11, they suffered their most serious stampede, and although no humans were seriously injured, more than 250 head of stock were gone. As Lorton had said on May 29, "To kill our cattle is to take our life." It occurred several miles east of today's town of North Platte, Nebraska, in the Maxwell or Brady area, and the train was delayed for six days while they searched for the missing cattle. Cephas Arms recorded they found 127 head as far away as fifteen miles the next day (June 11, 1849, in Cumming 1985, 13). "A few of the Jayhawkers swam across the Platte when they recognized their cattle [in a company of Tennesseans] including one of Mr. Doty's oxen which was white. When the rival train refused to give up the cattle... [Doty] asked the leader of the train if an oxen's word would be identity enough and the leader agreed. The oxen spoken to recognized his master and the herd was turned over" (H. Doty 1938, 2).

Sand was a distinctive feature in their travels—the bluffs and rock formations, but most particularly the difficulty of traveling through it. The toll it took on the animals pulling heavily laden wagons was considerable; however, it did not deter them from stampeding. "Travel over a heavy sandy road in the morning.... travel thro' sand & sand bluffs all day, some times 1/2 way up to the hubs.... Camp in sand bluffs" (Lorton, June 4, 13, 18). Part of the wagon road is still visible over the highlands north of the river near Birdwood Creek, about eighteen miles northwest of North Platte.

Soon they entered the valley of the North Platte River. On June 19, "during the night a hurrycane springs up & slits our tent. at sun rise next morning (being camped on a marsh) millions of musquitoes make their appearance to the great anoyance of all hands.... Buffalo came down off the bluffs & feed with our cattle.... being fenced in, they were soon dispatched by bullets. cut off some steaks for supper" (Lorton).

Accidents were common: "Boyd, while sitting on the [wagon] tongue, fell off & both wheels run over his back & leg—very badly hurt" (Lorton, June 19).[6] The injured and sick added more weight to the wagons and required special care. Cholera was still dogging the wagons on the southern side of the North Platte. "See grave opposite Ash Hollow & clothes of the dead man lying near. he died 2 days before with the cholera" (ibid.).[7]

The first distinctive rock edifices the Jayhawkers saw were the Ancient Bluff Ruins, or Castle Bluffs. "The bluffs to day are rockey & singularly formed. Some peek in form of a cone up towards the sky to a prodigious hight.... other[s]

FIGURE 2.4. Jail and Courthouse Rocks with North Platte in foreground, looking south as Jayhawkers would have seen them.

present to the eye a long chain of broken down walls, fortifications, ramparts, decaying castles, moss covered turets, rewined [ruined] cittadels, fallen spires. I see 1 spire 50 ft high standing alone, & another a tall slim sugar loaf. These are composed of a petrified soft rock & looks as if it washed with the rain" (Lorton, June 22).[8]

They soon encountered Jail and Courthouse Rocks, rising about 400 feet above the valley floor. But the most famous rock formation was Chimney Rock. Merrill Mattes pointed out that it "was no ordinary landmark. More enthusiasm and more diary space was devoted to this particular landmark than to any other of the covered wagon migration" (1969, 378).

On June 23 the Knox County Company camped across the North Platte from Chimney Rock. "Halks have made nests in the top.... standing alone as it does it calls forth wonder & amazement" (Lorton). Four miles west of today's Bayard, this prominent landmark pierces the sky 480 feet above the valley floor.[9]

Farther on, Scotts Bluff stands 800 feet above the North Platte, where a hard capstone keeps it from eroding as fast as the surrounding landscape. The stone wonders were pleasing to the eye, but the emigrant trains continued to have problems: "iron bars, fine stoves, boxes of soap, carpets, locks, nails & boxes lay by the way thrown out.... have a visit from over the river. they had 4 cases of Cholera.... Threw away enough in camp to load 5 yoke" (Lorton, June 22, 24, 25, 1849).

The Knox County train arrived at the ferry near Fort Laramie on June 28. "Saddles lay arround, trunks, chests, bacon, &c.... the river is deep here & we

FIGURE 2.5. Chimney Rock in 2009.

ferry on an old scow, 1 wagon at a time, swim our cattle. pay $3.00 & ferry ourselves" (Lorton).

Fort Laramie was a major stop on the trails to Oregon, California, and Salt Lake City. It was named to honor a French Canadian fur trapper, Jacques LaRamée (or Laramé, as spelling varies), who was killed in 1821 in the general vicinity. In the early 1830s William Sublette and Robert Campbell built a log fort on the left bank of the Laramie River not far from its confluence with the North Platte and named it Fort Williams. In 1835 they sold it to Jim Bridger, William Fitzpatrick, and some other mountain men who, in turn, sold it to the American Fur Company in 1836. That year Fort Williams was visited by Narcissa Whitman and Elizabeth Spaulding, the first white women to pass that way (Mattes 1969, 481, 482).

In 1841 the American Fur Company had to update and modernize their fort when a rival one, Fort Platte, was built nearby. The Fur Company upgraded with adobe walls and renamed it Fort John. The rising tide of the western movement stopped at Fort John/Laramie over the years—the Bidwell-Bartelson party, the first Oregon emigrants, John Frémont, Colonel Kearny, the Donner Party, Brigham Young, the Mormon migration, and the gold seekers. Over time, common parlance shifted the fort's name to Fort Laramie (Hafen and Young 1984, 70; Morgan 1959, 7). "Many wounded men [from emigrant companies] were left at the Fort in care of the troop[s], our comp. talked of leaving Gordon that got run over with the wagon" (Lorton, June 19, 1849).

As they entered the Black Hills of Wyoming (not the famous Black Hills of

FIGURE 2.6. Laramie Peak and Black Hills, Wyoming. On "a clear day it can be seen 100 miles" (Lorton, about July 1, 1849).

South Dakota), the country began to change. "Road gravel, hard, & hilly.... to the west [southwest] rises tall Larimie Peek, black & blue.... quarrel between the officers, capt. & pilot. dissatisfaction about the camp'g & with the pilot" (Lorton, July 1).[10]

> [July 3] up rocks, steep gravel hills, down vallies, ravines, & decent[s]. hard day on cattle. every thing covered with dust. no graves to day. 4 dead oxen, & 1 lame, 1 by the road. [mule train] frighten Capt Haynes teams.... cattles feet tender.... Camp on La Parele river [La Prele Creek].... drive the cattle 2 miles on[to] scant feed, have guard all night.... party of black snake Inds camped up in the hills. Inds keep away from the road, afraid of the cholera. cold night—almost friezes.... there is som[e] large trees on the La Parele. just above, it passes under a bluff, forming a subteraneous passage. (Lorton)

Cephas Arms (also in the Knox Company train) described this "passage" as a "natural curiosity in the shape of a natural bridge.... We have christened it Welch's Bridge in honor of one of our company [Adonijah Welch of the Rovers]" (Cumming 1985, 49–50). Today it is called Ayres Natural Bridge.

Adonijah Strong Welch became one of the most influential educators in American history. He was twenty-eight years old when he went to California with the Fayette, Michigan, Rovers as their designated geologist. Before his trip he had already received his bachelor's degree from the University of Michigan at Ann Arbor, acted as "principal of the preparatory department," studied law and passed the bar, and was principal of the Jonesville High School (Putnam 1899, 140; Hermann 1959, 61). Welch may have gone to California to improve his physical health: "Welch is in good health and spirits and is a first rate man for the expedition," wrote Henry Baxter, captain of the Rovers (in Cumming 1985, 47). Cephas Arms wrote, "I am very well pleased with that part of our company which are from Michigan. They are quite gentlemen in their intercourse and intelligent men.... one [Welch] has been principal in an Academy" (ibid., 42).

After his return from California, Welch started the normal school (teachers' college) in Michigan, and later, as U.S. senator from Florida, he espoused equal pay for women. In 1869 he became the first president of Iowa Agricultural College, today's Iowa State University, where he put those words into action. He seemed to do everything, but his prodigious mental energy often outstripped his physical energy (J. and L. Johnson 2017).

On the eve of the Fourth of July, the company had "a grand jubele.... load an old musket heavy, set a slow match & burst her" (Lorton, July 3). On July 4, Lorton wrote:

> The birth day of our american independence.... road tremendous dusty up hill & down dale, [a]round bluffs, hills &c. The dust has a curious & obnoxious smell made still more so by the mixture of manure. at dinner time we stop on a creek of clear water [Box Elder].... pritty gloomey prospects, but we are determined to go on till we cant go any further.
>
> near night we strike the Platte [5 miles southeast of present Glenrock,

FIGURE 2.7. Today's Ayres Bridge over La Prele Creek southwest of Douglas, Wyoming, was named Welch's Bridge by Cephas Arms for A. S. Welch, a member of the Knox County Company in July 1849.

Wyoming] after an absence [from the river] of 80 miles, & we find ourselves over the Black Hills. over the whole dist' it is destitute of grass with
the exception of little spots found between the tops of the hills, tho' nearly
all of the surface is covered with wild sage....

a mormon that owns 1 of the ferries says he has crossed 16,00 [1,600]
teams sence May at $3.00 a team, & he has bot $900.00 worth of provisions for the Mormons from emegrants. he came from salt lake this
spring & now wants to sell out.... there has been 9 men drowned at this
ferry this season. They cross here on a/c of grass on the other side, &
above cross back.

we camp on Deer creek [at present Glenrock, Wyoming], on some fine
grass that had been eaten off close. this is a most beautiful place, clear
pure water murmuring over the fine gravel bed, the tall & stately trees,
the green bushes, the birds singing sweetly.[11]

Many of these large cottonwood trees were felled in 1849 to make ferries to carry
emigrant wagons across the river.

Lorton then describes the Fourth of July celebration on the banks of Deer
Creek and the North Platte:

In the eve'g of 4th the Pioneers are heard blazing away for dear life, &
numerous other comp.... We then fired a feu de joy by the whole comp.
Then we had a speach by prof. [Adonijah] Welch. He stood on a box &
addressed us in a very eloquent manner & when he had done [we gave]
3 chears & another fue. All the while the welkin [heavens] rang with the
shout of musketry from the other camps. We then loaded & fired a volley.
The [Iowa] Rangers then marched up, & turning an angle near us fired as
they turned sepparately. We were then collected in a circle & saluted them
with 3 hearty cheers & fired a salute. We then went into a committee of
the whole to invite the Rangers to join in the eve'g jollification.... We then
proceeded to the Oscaloosa & Batavia comp. & were joined by them. In all
we had several hundred all in our lines, fireing round after round, with a
sentiment in between.... In this salute one of the Rovers [Peter P. Acker]
got a thumb shot off, [his gun firing] premature[ly]. The fireing kept up
till a late hour, & then the violin was brot out & a cotillion commenced on
the green. We smoked cigars & drank lemon syrup & [ate] antelope meat.
Our fourth dinner was rice & milk.

Many men lost their lives crossing the North Platte. Food, baggage, equipment, wagons, and wheels were damaged or totally lost. Oxen and mules

drowned. Time waiting for a ferry was spent reorganizing baggage, sorting, washing, mending, shoeing animals, and taking stock into the hills to find forage. Others hewed long cottonwood logs into canoes they lashed together to create yet another ferry. Wet and sandy ropes cut blisters into men's hands as they hauled the rafts across the river; others spent many hours in cold water pushing, shoving, and loading rafts and herding stock across the river. It was hard and dangerous work.

The Knox County Company crossed the North Platte near its confluence with Deer Creek rather than going upriver to the Mormon ferries at today's Casper, Wyoming. Their ferry was "composed of 4 log canoes fastened to gether," which they sold to one of the Bartholomew brothers, probably Luzerne, for half price— fifteen dollars (Lorton, July 7).

Lorton wrote the next day: "The dust blowing on your lips makes them sore.... see mule laying [be]side the road sick from drinking alkali.... I stripped, jumped in, swam backwards down the swift current, floundered about a little, washed, tried to swim up current but could'nt." The Jayhawkers probably joined Lorton for a swim before leaving the North Platte for the dry haul to Independence Rock and the Sweetwater River.

On July 12 Lorton described another natural wonder still unique along Wyoming county road 319 (called Oregon Trail Road): "The road passes thro' 'Rockey Passage' [today's Rock Avenue] running on both sides of the road.... The north wall runs some 5 hundred feet nearly in a straight line. The top runs from 5 to 50 ft high... The outter surface is smothe & perpendicular, assending & de-

FIGURE 2.8. Avenue of Rocks west of Casper, Wyoming. "The rocks form a complete wall, or passage running on both sides of the road tho' the south wall is not so perfect" (Lorton, July 12).

FIGURE 2.9. Willow Springs was the most welcome camp between the Platte and Sweetwater Rivers. For early emigrants, limited pasture surrounded the fresh-water spring.

cending hills in its course.... after dark we hall up at willow springs where we could hardly find a place to set tent, for the thrown away rubbish & piles of bacon."

Two days later he says they "come to Independence Rock. The rock rises perpendicular on the river side with enormous craggs. large loose rocks hang ready to fall & crush the passer. deep seams & fissures run through this enormous mass of stone, while the deep crivaces, chissled up & painted with hundreds of names, & the outer surface covered with 1000s."

There were significant points along the route, such as Forts Kearny and Laramie, that defined progress on the arduous road from the Missouri River to California, but Independence Rock was an even bigger goal—a psychological goal with a date attached. To reach Independence Rock by the Fourth of July was a milestone that signified a successful crossing of the Sierra Nevada before snow blocked the passes. Those who did not arrive at Independence Rock by the Fourth had the fear of failure added to their emotional burdens.

Independence Rock was also known as the "Great Register of the Desert." No one knows the total number of names, initials, and dates inscribed or painted on this anomalous granite monolith, but in 1999 Levida Hileman inventoried "2,056 entries" (2001, 116).

Beyond Independence Rock, there were two more goals, or points of progress, before reaching the goldfields via Salt Lake City—South Pass and Fort Bridger.

The travelers were now on the Sweetwater River. Even the name sings in the soul. It promises good water and grass all the way to the roof of the continent.

Just west of Independence Rock is Devils Gate, through which the Sweetwater flows. Its height above the river depends on where one measures—at the rim of the chasm or at the highest cliff point on the right bank of the river. The Bureau of Land Management says it's 361 feet, but the numbers have varied historically.[12]

Dow Stephens wrote about walking through, not around, Devils Gate:

> A party of us started, but there was only two to complete the trip, one other fellow, who nearly lost his life, and myself.... In some places we had to climb almost perpendicular walls, almost a hundred feet in height, then walk along a narrow ledge where a mountain goat would hardly venture. I have heard of foolhardy escapades, and have often wondered how we ever managed to come through with our lives, but luck must have been with us, for it makes me shudder even now to think of the danger we were constantly in. (1916, 12)

FIGURE 2.10. Devils Gate looking northeast. Dow Stephens said, "In some places we had to climb almost perpendicular walls,...then walk along a narrow ledge where a mountain goat would hardly venture. I have... often wondered how we ever managed to come through with our lives."

Excerpts from Lorton's journal highlight travel along the Sweetwater. Far to the north they could see the Wind River Range; close at hand were sage-covered hills and willows along the river. "Go in swimming quite often.... lots of antelope; Cooke cruller[s] & doenuts.... lighten our load.... have a good sing in the eve'g.... pass ice spring 1/4 mile from the rode...almost knock sage hens down with cane.... pass a rock perched on a bluff, formed like the walls of a house partly broken down, & hollow inside.... kill several antelope, prairie squirrels, badgers, wolves &c." (July 17–19).[13]

Yet another stampede occurred on July 19:

Capt Talors little boy was run over by waggon & oxen, another had a hand run over, another knocked in the head, another run over the leg, others knocked down, cattle became entangled in chains and wheels, axels broke, wheels [broke], wheels passed over oxen, & 1 ox was dragged way down the hill under the box with feet up. he had a horn broke off, & otherwise injured. broken horns lay around, hoofs broken.... trunks fell out, tubs pans & chains broke.

[It was] good for Capt Haynes the cattle broke the chain, [for] he was [still] sick. a team behind us would have smashed our waggon if Bartw [Edward or Luzerne Bartholomew] had not frightened them off & thus saved his own life & ours perhaps. They run on every side of the hill. some [had] on shues & pitched [them] off. They ran some 1 mile, & others to the bottom of the hill. such dashing pel-mel down such a steep desent is a freightful sight to gaze on when human life is in jeopardy, to say nothing of propperty, & that not more damage is not done & lives sacrificed. I can only account for it, That an all wise Providence looks down on the Knox Co. Comp. with a forbearing eye (Lorton). [14]

Tempers were short, and men were sore, fatigued, angry, frustrated, and anxious. Lorton "had a fight with E[dward] Bartholomew. he told me I lied, & [I] give it back. he struck me as I lay down. I jumped up & mashed his mug, made the claret fly & tore the shirt right off him, till he thought he wouldn't lick me any longer" (July 20).

Harrison Frans started a letter to his parents as the train lay by repairing wagons and searching for stampeded oxen:

Friends, wherever you are, I can say among the many trials and troubles and the many deaths caused by cholera, I have not had one sick day since I left home. Out of one hundred fifty men, there has been but one died;

but some are very sick now. Captain Hains [Asa Haynes] is very sick with the fever.

> We are camped at this place on the Sweet Water, trying to find some 20 head of cattle that took a stampede at 8 o'clock in the evening. There were but 35 wagons in the company, all huddled together—cattle [tied] fast to the wagons—men eating dinner—lying under wagons, resting— and some asleep;—all perfectly still. A wagon tung fell down, and before any man could get on his feet, the whole train was in motion. Every man was run over, and some seriously hurt, but no bones broken. I got a little hurt, myself. The cattle are just as wild as buffaloes, and can run a fast horse to death. (July 21, 1849)

Lorton also commented on the captain's health: "Capt Haynes, [afflicted] with sickness, aided by the numerous obstacles has, for a few days back, bordered on lunacy. his countenance looks haggard & his eye wild. he rode down the steep desent at our hill stampeed. The numerous dangers, the delays, the recollections of home, swell the burden on his troubled mind. H[aynes] was crazey & would go down to the spring to get water in his canteen, & drew his knife on those that tried to stop him. He was hurt in the hill 'stampeed.'"

Surely the Jayhawkers felt as discouraged as Lorton:

> Things now look gloomey.... we have be[e]n arrested in our progress, we have been delayed beyond forbearance, we have sustained loss after loss, we have been plundered of our property, run over by heavy ladened wag-ons, trampled beneath the feet of hundreds of infuriated & affreighted anim[a]lls, swam rappid currents under penalty of drounding, suffered fatigues, privations, thirst & hunger, ran under the tomehalk & scalping knife of the blood thirsty forrest fiend, "caraled" in the sand with no blan-ket & wild beasts preying & howling a "requiem" around your pillow of sod. difficulty after difficulty & obsticle after obsticle has presented itself, but from time to time they have been surmounted, & the jovial song has made the "welkin" ring, as they wound their way o'er hill & plain. but now misfortunes to[o] frequently occur, & seam not likely to terminate. things that before assumed a dark aspect, now begin to look black & serious. (Lorton, July 21)

But their resilience was still evident: "The nights are cold as midwinter but the days are hot. men sent out to south pass & the back [track] to hunt for mule & 3 ponies. I find them in the bluffs. cattle all brot in but 5. great rejoicing. I sing negro melodies & play violin" (ibid.) This is the first mention of Lorton's ability to play the violin. There are other comments about a violin being played, but it had not been clear that Lorton was the player. He was also noted for his fine singing.

John Colton's letter to friends written July 17–24 is a valuable review of their trip from the North Platte crossing to South Pass:

Camp on the banks of the Sweet Water, July 17, 1849.

Dear Friends, . . . We ferried the old Platte for the last time the day after I last wrote you. There were nearly fifty men drowned within fifteen miles of us in one week. I do not know how many drowned in it this year. It has the swiftest current of any river in North America, running 8 1/2 miles per hour. You can see by this, too, what a gradual rise we have to ascend. There has been no one drowned in our company yet, though Ward from Olmstead's mill came very near it. He was taken with the cramp in the current, but managed to keep his head above water until he struck a sand bar, when he was taken out.[15]

We have been following the bank of the river, so that water was handy; but we have traveled five hundred miles without any rain, and the sand is as hot as an oven. We comfort ourselves however, with the idea that we shall have none for five hundred miles to come.

It is forty miles from the place where we leave the Platte to the Sweet Water. There are two springs between, but no feed of any consequence. We had to carry hay. At last we came in sight of the beautiful Sweet Water, which was a welcome sight to us and our cattle. We first touched the river at the great "Independence rock," the spot where we should have been on the fourth of July. This is a great curiosity. It is one solid rock, rising abruptly in the centre of a beautiful valley. It is half a mile long by a quarter wide, flat on the top. I should think there were half a million of names on it at least, put on with all kinds of paint and brushes. I put mine on with tar and a rag under old Mr. Strong's name, who passed some time before.

Between the Black Hills and this place, a distance of 200 miles, we saw at least nine hundred dead oxen. The emigrants that lost them kept no guard over them, and being dry, they would drink from lakes and springs that abound in this country, which are poison, some of them consisting mostly of saleratus, and called Alkali lakes. The oxen drink it, and the first notice the teamster has of it, his ox drops dead in the yoke. A great many have lost their teams entirely, and are packing through on foot. We have kept a very strict guard, and have only lost two of our train. I have put my stuff into Urban Davidson's [Davison's] wagon, with one yoke of oxen, amounting to 450 lbs. We had to cut our [wagon] box, so that it would hold enough only for two. Most that have not already done so, are cutting off their boxes and projections, and throwing away things that they cannot carry....

Just as we passed Independence Rock, a few loose cattle from another train ran in among ours, and we then had one of the stampedes you read of. They all piled in together, wagons and oxen. No damage was done except a few men were badly scared. We were detained one day this week by our cattle taking a stampede in the night. We got them all again the next day.[16]

The emigrants have a new obstacle presented to them since they left Fort Laramie. About 300 troops have deserted with mules and arms, and riding along very pleasantly. When they get out of provisions, they go up to a train of three or four wagons, quietly present their revolvers, and politely request the teamster to hand over provisions. They then ride on till they are out again.

We find a great many natural curiosities in traveling through this country. By digging down one foot near a spring, we found ice ten inches thick, and near the Alkali lakes, I have seen better saleratus than I ever saw before. Soda, tartaric acid, saltpetre and salt are all found. A large lump of gum gu[a]iac was brought down from the top of the Sweet Water mountains the other night. The guide says we shall be in a few days where honey drops out of a solid rock. This is his word for it. He also has in his possession first rate rifle powder, which he procured by boiling down water from a spring in the Rocky Mountains.[17]

We are now following the Sweet Water. A great many cattle are left, given out entirely. If mine give out I shall make a knapsack.
A few days later Colton found time to continue his long letter home:

July 20. I will try and write you a few more lines this afternoon, as we are lying by. On the 18th we traveled eight miles, and stopped at noon near some ice springs on the top of a hill. Some of the company took their oxen off the wagons, others left them on. It was very warm and we were all lying under our wagons, when all of a sudden the cattle started! Those hitched to the wagon rushed upon each other, running over men, women and children. Our oxen were off the wagon and hitched together. I was lying under the forward axletree. Harry Frans' wheel struck ours and broke the axle. Everything was in perfect confusion; one yoke of our oxen got hitched under Frans' wagon and were dragged ten yards on their backs before we could get them loose, and then they were nearly dead. I don't think we can use them again. A train passed us just after this occurrence, and I bought an axle for $1.00, and hired a wagoner of the same train to put it in. The balance of the train went on twelve miles to camp. We got ready to start about sundown. We traveled about ten miles after dark, but could not find the camp, so unyoked and camped alone. We tied our oxen to our wagons, without feed since morning, and had not been abed half an hour, before about 20 head of cattle came running by as hard as they could; ours tried to get away but couldn't. Next morning we went on to the camp but found them all in disorder. Their oxen the night before had stampeded, and ran in all directions; it took three days to find them; so we concluded that our broken axle had brought us good luck at last. It is said there has never before been a stampede on the Sweet Water. We must be an exception. We go faster than any other train, but our cattle are so well used that they run every chance they got, whether they are tired or no. They hear noises that frighten them, when we can hear nothing; and when we hear something that we are sure will start them, they mind nothing about it—it is a perfect mystery. We have heard from [about?] one yoke of the oxen which we lost, and have sent back after them.

Things are cheap here. I bought 20 pounds of coffee the other day for 50 cts. Large quantities of provisions are left by the way, which cannot be hauled. Since I bought that axle I have baked bread by [burning] wagon wheels—they make excellent wood, they are so well seasoned.

We have divided our company into small ones on account of Stampedes; no company should consist of over ten wagons. During the three days that we have been here, three oxen have died by drinking alkali. We are twenty miles from South Pass [camped at Strawberry Creek]. The nights are very cold; water freezes. Every day counts as two now, for we are in danger of being caught in the snow in the Sierra Nevada. There is but 12 hours' danger, but in that time it is either life or death. One hour it is sunshine; the next, snow.

Four days later Colton and company crossed their long-anticipated goal—the continental divide. John again finds time to take pencil to paper:

July 24. We are lying at the Pacific Springs in the South Pass. We have traveled only eight miles to-day, on account of having to go without food or water during the next 24 hours. We have separated ourselves to-day from the eastern side of North America, nearly 8000 feet above the level

FIGURE 2.11. Around Pacific Springs the teetery, sloppy, grassy tufts look stable, but they float on soggy, unstable mud, as LeRoy discovered.

of the sea [7,411 feet elevation]. We reached the summit at noon; we could see the Sweet Water running one way, and Pacific Creek the other; after this all the streams run into the great basin called St. Mary's sink, and then disappear.[18]

We are now 200 miles from Salt Lake, and will be there in two weeks. We have been disappointed about meeting the mail, but Babbit had had bad luck; one of his men died of cholera, and he had to lie by. He calculated to have made the quickest trip ever known.

We have had no more stampedes, and I think we shall have none. Our teams look well.

Last night a Mormon [John D. Lee] with his team camped near us. He is from Salt Lake, and is going to help one of his brethren on the way. He says crops are fine there, and companies are coming back from the diggings with mule-loads of gold money—money as plenty as dust! That sounds a little encouraging to us poor fellows, who came near freezing last night.

To-night is the last of my writing in this letter, as the Mormon said Babbit would be along to-morrow. We have got over the worst, and have good feed and water. As to advising people to go to California, I would say, go if you want to.

Yours, Truly, J. B. Colton

Harrison Frans continued his letter at South Pass:

This day I have gone through South Pass, but it was a tight fit [a facetious statement]....[19]

This will be a great year for prodigal sons,—many are returning home.

Though the roads are better than what you have in the states, common two-horse wagons are heavy enough for the trip; five yoke of oxen, well taken care of, will perform it. I would advise every man to stay at home or go by water, for it is almost impossible to get feed. Sometimes [we] drive fifteen miles from the road to get feed, the grass on the road being entirely eaten off.

I do not want any of my friends to attempt this trip. There are plenty of people to suffer this season, for they cannot get through. There are some ten or twelve thousand teams on the road at this time....

There is no timber on this road after you leave the Missouri river, unless it is cotton-wood or willow. There is not fifty acres of land in this territory that you could raise anything on.

Tell the people I have not gone wild yet, but follow buffaloes two days and then come back.... It is no use to write me, for I will never get the letter, the distance is so great. (July 21, 1849)

John Groscup (1885) described their travels succinctly with his creative spelling: "wee toiled threw Mud rain & wauter at the start, threw thundur & lightning and rain [and] hail Along the Plat, & indians & Buffalo Antelope Along the plaines un till wee came to the black Hills bare whare Meeney A wagon was left in dis spair. shugar & rice in barrels ware left.... all Along this road with spoils did lay but still wee went on with Curage bold untill the City wee did be hold and then wee listened to talk untrue and laid us by some Months ore to."

Alonzo Clay wrote back home from Salt Lake City on September 22, 1849:

FIGURE 2.12. Plume Rocks, now called Bluff Rocks, looking north.

You cannot form the most distant idea of the amount of suffering that some of the emigration have passed through this season enroute to the mines. Cattle have stampeded and been lost. They have died in the yoke in the road caused by hard driving & the Alkali on the grass until waggons have had to be left by the road side and the owners obliged to take a pack on their backs and go on foot to the mines a distance of 800 or 1000 miles. Add to this the [*tear*] that must be caught in the mountains by the snow by the trump[?] of the season. Look at all this and you can form some estimate of the suffering that must be endured to get that treasure "which perisheth."

If you or any of our friends conclude to try it, I would advise them to take hor[s]e teams or mules and put them before light waggons with just provisions and clothing enough to last to the mines. Take three or four good horses or mules to each person and start from the Missouri River by the 1st day of May with plenty of money in your pocket and you can get through safe enough, but do not take more than 800 or 1000 lbs to a waggon and 2 men—Women have no business on this road if they ever wish to enjoy any degree of comfort.

South Pass to Salt Lake City

As the Jayhawkers set out on the western slope of the Rocky Mountains on July 25, Dow Stephens, future Death Valley Jayhawker, entered Salt Lake City with others from Fulton County, Illinois. The Jayhawkers started one week behind Stephens from the Missouri River and lost another week primarily to damaging stampedes.

Between South Pass and Green River three things along the trail were noteworthy: a strange rock formation, the cutoff to Fort Hall, and the death of a comrade.

About nine miles westward from Pacific Springs, the trail passed south of Plume Rocks, a knob from the Wasatch Formation of variegated mudstone and sandstone. Lorton called it an "eastern lamp or pitcher," an Aladdin's lamp. Goldsborough Bruff described the landmark on August 3, 1849, as "some low clay bluffs, of a dark dingy red hue, and singularly plume-formed projections on top, from the effects of elements" (1944, 1:65).

They crossed Dry Sandy Creek where the water was brackish and not good for cattle. Six miles farther was a major fork in the Oregon-California Trail now called Parting of the Ways. The left fork went to Fort Bridger and Great Salt Lake; the right fork, commonly called the Sublette Cutoff, went to Fort Hall in today's Idaho.

Here they found a notice that J. R. Nelson was not likely to live. They crossed Little Sandy Creek and camped on the Big Sandy River, where they found feed for the cattle.

FIGURE 2.13. Green River at California Trail crossing above the confluence of Big Sandy and Green rivers, looking upriver.

On July 27th, Lorton gave a description of a respectful burial thousands of miles from civilization:

> We burry Nelson. put on his clothes, wrap him in two blankets, a pillow under his head, & then pall bearers take hold of the blanket on each side followed by the rest of persons present, [and] carry him to his last resting place. the grave was upon the edge of the bluff, the usual depth, while the bottom being hard earth was dug out in shape of a coffin. in here he was deposited. boards were laid crossways, & upon these were laid willow branches & the earth throw[n] in upon his last remains. The head board [was] placed & the grave rased, in due form. a tall man stood by which was the only person who shed tears & his were principally on a/c of the deceased haveing no coffin. he said he took all his boxes & chests & spliced them to build a coffin for a little girl that died in his comp. Thus we left him with no useless coffin enclosing his head, upon the banks of Big Sandy.[20]

Nelson is not listed on the company roster, but Lorton made a headboard for the deceased reading, "J. R. Nelson. Died July 26th 1849 Adams Co. & a Member of the Knox Co Comp."

After a fatiguing tramp on blistered feet, they arrived at Green River about noon. Lorton says, "This is a mammoth river for a mountain stream, about 100 ft wide & rappid current. The water looks green." There was a Mormon-owned ferry that cost four dollars per wagon, but the Knox County Company traveled to a ford twelve miles downriver.

Sheldon Young's party took a different route from Green River into Fort Bridger and had already arrived in Salt Lake City on July 22 while the Jayhawkers were approaching South Pass. Sheldon later became a Death Valley Jayhawker.

The various company divisions followed down the Green to Blacks Fork, where all the emigrants experienced "plenty of musquitoes" and hot sun. Lorton also found wild currants and made sauce from them. Next to the trail they saw a "formation of sand stone (brown) with all the carvings of an artist, every variety of dome from a Turkish mosque to a St. Peters dome"—today's Church Butte (July 30).

Lorton's division was no longer with the Jayhawkers, who were now a day behind. The latter arrived at Fort Bridger on August 1. Around the fort they found the valley

> litterally covered with squaws, naked papposes, & indians on horseback, some riding at full spead, & others lying & lolling on their ponies backs. here there is some 1/2 dozen lodges formed by long poles 15 ft high cross-ing at the top to let smoke out. besides the buffalo hide that form the roof is opened & kept so by long poles. the skins are sewed together with sinues.... these traders lodges are filled with goods, a great deal bot from Cal[ifornians]. (Lorton, July 31)[21]

Bridger's fort "is made of logs, or rather the cabins form the fort, & the yard is the middle, a large plank gate way to the entranse. on the north end is the store con-taining good[s] for traffic.... The roof of the fort & chimnies are made of sod" (ibid.).

After leaving this important landmark, they traveled west to Soda Hollow (well south of Interstate 80), where they found soda water, which "is rather strong, & stings in your nose" (Lorton, August 1). They ascended high mountains and down "awful gulfs, pass little trees, berry bushes, over rockey steeps," until they came to Bear River, about eight miles south of today's Evanston, Wyoming.

The Jayhawkers continued west past Cache Cave, "where there is a fine spring of water," and entered Echo Canyon (merging with today's Interstate 80) to Weber River. They were now about forty miles northeast of Salt Lake City. All along this area they had seen Indians (Lorton, August 4, 1849).[22]

Now began the difficult crossing of the rugged Wasatch Range. They turned down the Weber River and passed the rock formation now called the Witches, where Lorton paused long enough to make a sketch of them. At today's Henefer, Utah, they turned south into Main Canyon, Dixie Hollow, and into part of East Canyon. The track took a sharp right turn to the west and entered Little Emigration Canyon,

FIGURE 2.14. Reconstruction of Bridger's Fort, Wyoming Historic Site, on Blacks Fork off Interstate 80 near Lyman, southwest Wyoming.

where they camped "at a spring in the mountain." This is the road the Donner Party blazed in 1846. They crossed Big Mountain Pass, elevation 7,420 feet, and started the steep descent into Salt Lake Valley where they arrived at the city August 8, 1849.

In the Valley of the Saints

Civilization brought gratification of two cravings held by the young gold seekers: the sight and company of young women and the desire to eat fresh vegetables! Lorton mentioned on August 22, "have singing at sister Pratts & pritty girls," and Charles Mecum reminisced in 1878: "How the time passed gaily till we arrived at Utah: of our rest in that place—while we danced with the fair lambs of the flock of Brigham. I remember our division:—part going the *northern*,—part the *southern* route. Ah! could we *then* have foreseen the *future*—could we have *then* known the terrible ordeal through which we must pass."

Of his craving for fresh vegetables, Lorton said:

I ate & ate as if I would never stop eating green vegetables. to set up to a table seamed awkward & to lean back in a chair a novelty. the log cabin in which I ate had a homely look, & it seamed altogether as if I had got on board of civilization once more. I would take up a piece of hot corn, select out three rows for my corn crackers & go to work in my own celebrated stile. then up come the peas, then the beans, & then the beets. my eyes were very large, as could be seen from the quantity on my plate. by & by they all got done, & I only just began. however, I was not to be backed off, bu[t] kept on till another table full sat up. (August 15)

Many of the men found employment working for the Mormons during their stay in Salt Lake Valley. Mr. Erkson, bound for California with his wife, remembered:

> Several of us went to work getting out lumber for Brigham Young while we were waiting and resting. The Mormons all advised us not to undertake to go on by the northern route, and as the travelers gathered at this point they canvassed the situation. We used our teams when we were at work for Brigham and assisted in building a dam across a cañon where he intended to build a woolen mill. I earned about a hundred dollars by my work, which was paid to me in ten-dollar pieces of a gold coin made by the Mormons. They were not like the U.S. coins. I remember one side had an eye and the words—"Holiness to the Lord." (in Manly 1894, 367–68)[23]

The Mormon Church started their coinage system in late 1849 and minted $2.50, $5.00, $10.00, and $20.00 gold coins.

Two years before, on July 24, 1847, a year after the Donner Party, Brigham Young and his entourage also exited the mountains through Emigration Canyon to begin the laborious process of establishing a New Zion for the Church of Jesus Christ of Latter-day Saints (Bullock 1997, 237). By August 1849, when the Jayhawkers arrived in the valley, the Mormon population was about five thousand, with another fifteen hundred on the way.

FIGURE 2.15. Witches Rocks on Interstate 84 (Weber River) near junction with I-80. One of several notable formations.

In order to build their kingdom in the West, the Mormons needed an easier and quicker connection to the states and Europe to replace the long, laborious wagon road across the continent that was closed during winter over the Rocky Mountains. They needed access to a seaport; thus, a wagon road southward to the Old Spanish Trail, thence to Los Angeles and the seaports at San Pedro and San Diego, was essential to their growth.[24] Interstate 15 now serves that purpose.

The Mormons knew such a road was possible because the route had been pioneered by their packers who traveled to Los Angeles in the winter of 1847–48 (Sutak 2010, 86). Captain Jefferson Hunt (and others) made the return trip in the spring of 1848 with a herd of cattle, and Orin Porter Rockwell, also one of the packers, returned two months later with men from the Mormon Battalion and one light wagon owned by Captain Daniel C. Davis (Bigler and Bagley 2000, 396–400).

How could the needed freight road be built quickly? Fortunately for the New Zion, an answer lay close at hand. If the Mormons could encourage the gold-seeking emigrants, who were recuperating in the Salt Lake basin, to travel the Southern Route, these Gentiles would have to build a wagon road to Los Angeles. These strong young men were anxious to get to the goldfields and were willing to pay for a guide to Southern California.

Circumstantial evidence indicates the Mormon hierarchy had a plan to entice the emigrants to travel the Southern Route. First, Brigham Young preached a "sermon that the Lord had come to him in a vision and told him that no emigrants starting after that time over the northern route to California would arrive there, but would leave their bones to bleach on the plains or in the mountains" (White 1927, 207–8). Second, stories of death and destruction on the Northern Route were circulated and embellished. Mormon Battalion veterans who returned to Salt Lake City from California gave eyewitness accounts of the horrid squalor found at the hastily built Donner Party log cabins (Jones in Bigler and Bagley 2000, 238). William White took the Salt Lake Cutoff to the California Trail (Northern Route), where he says he was "never out of sight or smell of dead cattle and mules and horses" (1927, 207–8). Stories were circulated about hundreds, if not thousands, of rotting horse, mule, and ox carcasses littering the trail along the Humboldt River trail. Goldsborough Bruff sketched one such ghastly scene (1944, 1:202). James Wasley wrote about these stories on October 21, 1849:

> Report said that the grass was all burnt off on "Mary's River," there was no feed left for teams, that cattle were dying off by hundreds! That women and children were sitting by the road side, barefoot and crying for bread, that the men formed themselves into gangs, and robbed the trains as they came up, that there would be snow on the mountains before we could possibly reach there,

and we should perhaps perish in the snow.... They recalled to our recollection the fate of the "Donner party," and also of Fremont and his men last winter.

The third part of the Mormon plan was the threat that not enough food would be available for emigrants who remained in the valley over the winter. Although the *Kanesville Frontier Guardian* proudly announced on August 8, 1849, "The wheat crop [in Salt Lake Valley] has come in bountifully, ... [and there was] every prospect of a large surplus of food for man and beast," Brigham Young reported in October: "Several hundred of the emigrants arrived in the Valley too late in the season to continue their journey on the northern route, and many of them contemplated wintering with us. So large an accession of mouths in addition to those of our own emigration, threatened almost a famine for bread" (*Journal History*, October 21, 1849).

The fourth part of the plan was to offer a Mormon guide to lead a train to Los Angeles (for a price). Captain Jefferson Hunt, who had traversed the Southern Route the previous year, was under direct orders from the Mormons' First Presidency (Brigham Young) to pioneer and measure a wagon road to the Pacific Ocean. To measure the road, a wooden roadometer the Mormons had perfected was attached to the wagon Addison Pratt was in (*Journal History*, September 20, October 2, 1849; Hafen and Hafen 1954, 224).[25]

Thus, the gold seekers were inundated with dire information and at the same time urged to move on. Lorton summed up the predicament on August 14: "Never was I in [such] a delema before, or ever before struck so forcibly that it was nescesary to do something, & that immediately.... & what made it worse, every body was in the same fix. No one knew what to do. To cross the trackless desert [via the Southern Route] was burring [burning] yourself alive, & to go the Northern rout rushing in to a grave yard & rideing over dead bodies."

Disagreements over the two routes wrenched apart messmates, close companions, and even brothers. Lorton reported August 15: "Part of our mess with others started for Cal. via north [route, with] 6 yoke of oxen, carr[y]ing 1 change & provisions just enough to last." Luzerne Bartholomew separated from his brother Edward and reached Sacramento November 3, 1849, via the Lassen route (Runner 1995, 2). Edward opted to take the Southern Route with his Jayhawker buddies, and he spent Christmas Day 1849 in Death Valley. He arrived at the goldfields almost four months after his brother. A. J. Baker of the Rovers also chose to go north, but most of the Rovers took the Southern Route (Cumming 1985, xi), as did many from the old Knox County Company.

Thus, the stage was set for the travails of the Southern Route and the loss, agony, and suffering yet to be endured by the Death Valley '49ers.

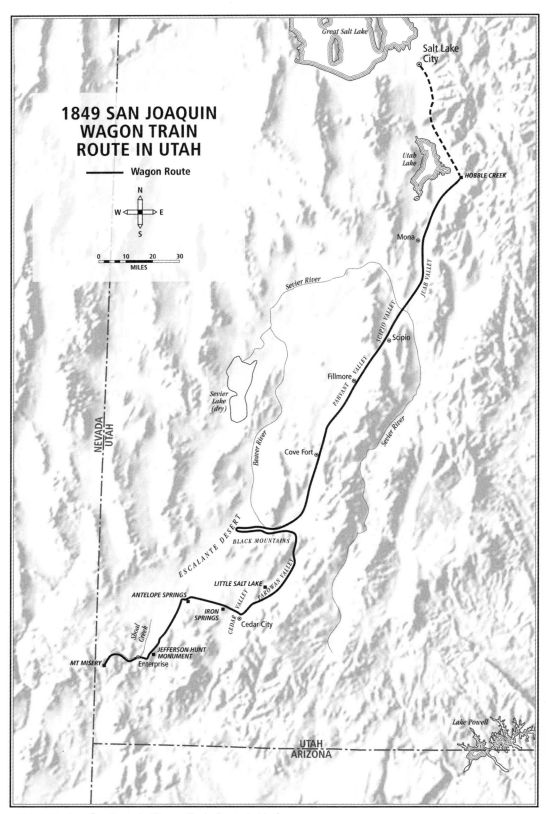

FIGURE 3.1. *1849 San Joaquin Wagon Train Route in Utah.*

CHAPTER 3

GREAT SALT LAKE CITY
TO MOUNT MISERY

JAYHAWKER ALONZO CLAY wrote home from Salt Lake City:

> We shall start on Monday next [September 24, 1849] by the way of
> Pueblo de los Angeles for the "promised land." We arrived here on the
> 8th of August and have been in this region of country ever since.... The
> Mormons are now engaged in building a council house forty feet square
> which is the only public work they have yet commenced.... Poligamy is
> common here. the more women a man can take care of here [the m]ore
> he is thought of.... There is evry reason to believe that gold is plenty here
> but it is a church secret for the present. (September 22, 1849)

The Platte Valley Jayhawkers opted to continue their journey by taking the
Southern Route, and they slowly made their way about sixty trail miles south
of Salt Lake City to Hobble Creek at today's Springville, Utah (about six miles
southeast of old Fort Utah at Provo). Others from the Knox County Company also
took the Southern Route—Cephas Arms, the Ewings, the Taylors, the Rovers (all
but one), and William Lorton.

The Southern Route

The Southern Route has been analyzed and described in a number of books, such
as Tom Sutak's *Into the Jaws of Hell*, Harold Steiner's *The Old Spanish Trail Across
the Mojave Desert*, Leo Lyman's *The Overland Journey from Utah to California*, and
C. Gregory Crampton and Steven K. Madsen's *In Search of the Spanish Trail*.
Sheldon Young's diary also covers much of the route from Salt Lake City into the
Escalante Desert west of today's Minersville, Utah. Then about two weeks are
missing from his diary (October 24 through November 8), as though he had mis-

placed it. It resumes just before he leaves Mount Misery to head west to become one of the Death Valley Jayhawkers.

The most complete firsthand accounts covering the wagon route between Salt Lake City and Mount Misery come from Adonijah Welch and William Lorton. As the official scribe for the San Joaquin Company wagon train, Welch's journal along with Lorton's are invaluable companions to Young's diary.

On September 24, 1849, we find the Jayhawkers drifting southward toward Hobble Creek to rendezvous with other wagons gathering to break a road down the Southern Route to California. The first night after leaving Salt Lake City, Lorton said he

> could see thro' the dark an ox team, & 3 [2] persons just came up & said they were lost & asked which way the road went. I then came up & found Davis & Clark the persons lost [Jayhawkers Urban Davison and Orrin Clarke].... after traveling a great ways we accidently fell in with [Edward] Bartholomew & Capt Haynes who had also lost them selves & camped here. they chained up there oxen to the bars & I posted my poney near some bushes to feed on them. (September 26, 1849)

The men wound their way through today's Sandy and Draper, past a boiling spring, across the tableland of the Traverse Mountains, and down to the Mormon fort on the Provo River, where "the Knox Co Comp. are camped," according to Lorton (September 26). He continued, "The fort here is built of logs, the doors inside & 2 gates, 1 at north & other south. an open log shed [is] built of huge logs & a heavy mounted cannon on top brot by Capt Hunt from Cal."[1]

A trial was held at the fort—Rovers v. Charles Dallas, who had several wagons of trade goods he was taking to California. The dispute was over some of the cattle the Knox County Company lost during the June 10 stampede, the ones Captain Baxter spent a week looking for when he temporarily left the company at Fort Laramie (Lorton, July 5).

Charles M. Dallas was the man who had hired William Lewis Manly and John Haney Rogers as drovers at the Missouri River. Seven of the crew decided to float down the Green and Colorado Rivers to California because Dallas planned to let them go in Salt Lake City when he suspected he would be forced to overwinter there. Instead, Dallas took the Southern Route, which left the river runners floating down the Green River somewhere in southeastern Utah. After the trial (the result was not recorded), they had "a 'Fandango' in the fort which lasts till day light" (Lorton, September 26).

By September 30, Sheldon Young had arrived at "hobble Creek the place of Starting · abundance of grass & good water weather continues warm & no rain."

It is time to get better acquainted with Sheldon Young, whose diary is so important to the Nevada-California part of the Jayhawker trek. Born in Connecticut on July 23, 1815, Sheldon was thirty-five years old when he struck out for the California goldfields; thus, he was more mature than most of the Jayhawkers. His obituary, August 25, 1892, said he had been a sailor but "gave up the sea for ship building, and in a chest...is a varied assortment of tools plied in his craft. Each and every tool was made by his own hand: he was a wonderful man in this respect." It went on to say he was a "very eccentric character" who "lived and died alone" two years after his wife's death (Young 2010, 59).

Young started his diary when he left Joliet, Illinois, for the goldfields on March 18, 1849. It ended December 15, 1850, when he crossed the Isthmus of Panama on his way home. He started to California with Gustavus Pierson and Hiram White. They shipped most of their goods by riverboat to St. Joseph, Missouri, while they traveled overland through Galesburg, Illinois, and Burlington, Iowa, to St. Joseph, where they retrieved their baggage. Thus, Sheldon traveled the same or similar roads to St. Joseph as did the Jayhawkers. However, he crossed the Missouri River a few miles north of St. Joseph and followed the St. Joe road to its connection with the Oregon-California Trail, then followed that trail on the south side of the Platte River. Somewhere in today's Nebraska, Sheldon mentions joining a company from Missouri and a man named Gritzner. By the time his mess headed into the Nevada-California wilderness, he was traveling with Frederick Gritzner, Wolfgang Tauber, and possibly William McGinnis. Colton included all but McGinnis as members of the Death Valley Jayhawkers (although he called Tauber "Wolf" and could not remember Young's name until 1897). (See chapter 10 for more about Young's life after 1849.)

The same day Young arrived at Hobble Creek, the inquisitive William Lorton met two other persons who are prominent in the Death Valley story, Lewis Manly and John Rogers, who were hiking out of the Wasatch Range.

> I see 5 men with 2 horses packed coming down the big kanyon to [from] the S. east. I went to meet them & learned they were Green River floaters that had left Charles Dallas, made canoes & floated down Green river to the Colerado. they said they had had a hard time of it & nearly all lost their lives dashing amonge the rocks & down rappids & over falls. they had all lost their fire arms except one, & would have starved to death but for Ind Walker who gave them as much as they could eat, & traded the horses to them for amunition & clothes, put them on a trail & told them it was 8 sleeps to the [Salt Lake] valley. the inds were very kind to them. they [the men] had hardly any clothing & shoes on them. they said it had

been very cold on the mounts, that they had seen mountains of every
shape & form. Walker [Ute chief Wakara], who is named after the great
mountaineer, gave them a map of the country in the sand. he would heap
up the sand for mountains, the valleys in between & mark out the roads
& rivers with a stick. (Lorton, September 30)

Wakara, born near Spanish Fork, Utah, became an influential Ute chief who
brought wealth to his tribe through extensive trading with Anglos. As typical of
an Indian chief, he was an excellent horseman and defender of his people. He
was considered a diplomat and spoke several languages. He saved the lives of
these emigrants by getting them to end their river adventure.

Lewis Manly, John Rogers, Charles and Joseph Hazelrig, Alfred Walton,
Richard Field, and M. S. McMahon had quit the employ of Charles Dallas when
they reached the Green River. The seven had repaired an abandoned ferry boat
and floated down the river, expecting to reach the Pacific Ocean by water. After a
tumultuous venture, they had a chance encounter with Chief Wakara where the
Old Spanish Trail crosses the river a couple miles upriver from today's town of
Green River, Utah. Field and McMahon stayed with the Indians and later trekked
through the Wasatch Mountains to Salt Lake City. The former five parted from
Wakara's band and were heading northwest toward Salt Lake City when they
came upon the gathering at Hobble Creek (Manly 1894, 74–103, 297–319; L. and
J. Johnson 1987, 51–52; Kane 2008, iv).

Lewis Manly wrote *Death Valley in '49*, one of the truly great adventure sto-
ries of the American West. Lewis and his Wisconsin trapping partner, Asahel
Bennett, planned to travel together from Wisconsin to California, but while
Manly went north to buy an Indian pony, Bennett started west with his family
(Sarah and their three young children) and took most of Manly's gear with him.
At Council Bluffs, Manly missed the Bennetts and hired on with Charles Dallas as
a teamster, but Dallas opted to take the Oregon and California Trail on the south
side of the Platte River instead of the Council Bluffs Road north of the Platte.
You can imagine Manly's surprise at finding the Bennett family encamped at
Hobble Creek! Here he was, nearly destitute, and he catches up with the Bennett
wagons that are carrying his whole "outfit and gun." Manly had left Wisconsin
with Bennett's "light gun, a small, light tent, a frying pan, a tin cup, one woolen
shirt and the clothes on my back" (Manly 1894, 61; Kane 2008, iv). After crossing
the plains and floating 430 miles down the turbulent Green River, he had almost
nothing left.[2]

The Bennetts had taken the Council Bluffs Road after joining the Badger
Company that Alexander Combs Erkson helped organize at Kanesville (Erkson

in Munro-Fraser 1881, 709; Manly 1894, 367). The Badger Company played hop-scotch with the Knox County train, and both Lorton and Cephas Arms mention the company. Several men (and families) from this company took the Southern Route toward Los Angeles.

Manly described his astonishing rendezvous with the Bennetts:

> We reached Hobble Creek before night.... There seemed to be no men about and we were looking about among the wagons for some one to in-quire of, when a woman came to the front of the last wagon and looked out at us, and to my surprise it was Mrs. Bennett, wife of the man I had been trying to overtake ever since my start on this long trip. Bennett had my entire outfit with him on this trip and was all the time wondering whether I would ever catch up with them. We stayed till the men came in with their cattle towards night, and Bennett was glad enough to see me, I assure you. We had a good substantial supper and then sat around the campfire nearly all night.... I had missed Bennett at the Missouri River.... After breakfast Mr. Bennett said to me:—"Now Lewis I want you to go with me; I have two wagons and two drivers and four yoke of good oxen and plenty of provisions. I have your outfit yet, your gun and ammunition and your two good hickory shirts which are just in time for your present needs. You need not do any work. You just look around and kill what game you can for us, and this will help as much as anything, you can do." I was, of course glad to accept this offer, and thanks to Mr. Bennett's kind care of my outfit, was better fixed then any of the other boys. (1894, 102, 106)

Bennett also invited John Rogers to join them, and they all became Death Valley '49ers. To assist the two Hazelrigs and Alfred Walton, Manly was able to procure "flour from Mr. Philips [a Wisconsin acquaintance] and bacon from some of the others, as much as we supposed the other boys would need, which I paid for, and when this was loaded on the two colts, Hazelrig started back alone to the boys in camp. As I was so well provided for, I gave him all my money for they might need some, and I did not" (ibid., 106).

On October 1, 1849, wagons began to move south to Spanish Fork and on to Peteetneet Creek at present Payson. Henry Baxter, captain of the Rovers, was voted colonel of the San Joaquin Company, which was organized with seven divisions, each to break trail and stand guard one day a week. Some division names were Jayhawkers, Bug Smashers, Buck Skins, Hawk Eyes, and Wolverines (Stephens 1916, 16). The Jayhawker division included the Platte Valley Jayhawkers plus other messes (or wagon groups) that were arbitrarily put into it. Snow was already falling in the mountains, and ice froze in the buckets. At first there were seventy wagons, five hundred oxen, and about one hundred horses feeding around the camp (Welch, October 3, in Ressler 1964, 258). Jefferson Hunt, a retired captain of the Mormon Battalion, was their pilot, to whom they paid ten dollars per wagon.

The company train encamped on Summit Creek, at present Santaquin, "a small swift running stream of the coldest water, four yards wide," on October 4 (Welch, ibid., 259), and then they continued to Punjun Spring, today's Burraston Ponds, southwest of Mona, Utah. Salt Creek at Nephi was next. On the Sabbath they "listened to three sermons. One at eleven by Rev. Mr. Ehlers a German reformed pr'er [preacher]. And then another at three by Rev. Mr. [James W.] Brier a methodist. And in the evening by Mr. Crow in the German language a portion of the com[pany] being germans" (Welch, October 7, ibid., 262). Louis Nusbaumer and other German-speaking men from New York had been in the California Mining Company. Nusbaumer and his messmates became Death Valley '49ers, as did Reverend Brier's family.

The train traveled down Juab Valley and camped near some "large excavations in the earth filled with water which from their singular resemblance to a cistern we have called Cistern Springs (Welch, October 8, ibid.). Next day they came to the Sevier River, where there were "plenty of the snake Indians about camp to trade an old flint lock gun will fetch a good Pony" (S. Young, October 9, 1849). They were Paiute. Lorton traded his "mare off for a beautiful Cal. horse, as white as alabaster. I gave a rifle to boot, 2 Boxes [of] caps, an old shirt &c." (October 9).

By October 10 Sheldon Young said, "we have now 100 waggons 400 men" at the Sevier River, the number Hunt had agreed to guide to Los Angeles. Next day they "lay in camp on accoun[t] of two Sick men not being able to travel." Lorton took advantage of the time to "do a large baking, 12 skillets full. In the eve'g have a negro [melodies] concert in N.Y. boys tent." Lorton felt comfortable with men from his hometown.

As the San Joaquin Company traveled west of south through the valleys bordering the western base of the mountainous backbone of Utah, Jayhawker Tom Shannon had two experiences that may have been in the Pahvant Valley area, eighty or ninety miles south of Salt Lake City:

> I made two discoveries on the desert, and do not know if any one shared them. the first was a well of warm water that ebbed and flowed. when I came first upon it, it seemed to be a well about 8 feet in diamater, perfectly round and smooth. it was eight or nine feet to the water. while I was contemplating how I could get some water both to drink and to wash, it

became agitated and slowly rose to within three feet of where I stood, so
that I put my hand into it and felt its temperature. it was right for a warm
bath. this flow was also acompanied by a roaring and gurgling sound,
and the earth slightly trembled under my feet. as I found the [water] undrinkable, I determoned to take a plunge. But when about ready the water
began sinking with a gurgling sound. I stood and saw it recede to where
it was at first. in ten minutes it came up again and I plunged in to take a
quick bath, but it went down before I could get out, so I had to wait the
next flood tide and it came, but I thought it was an eternity getting there.
The other discoverey was a kiln where once had been made and burned
the pottery, the fragments of which we saw so much. it was about 12 feet
in diamater, 8 feet deep, was built of coarse cement and burned red after
being built, acres of broken potery around it. (1895)

The well-known Meadow Valley hot spring, about thirteen miles southwest of
Fillmore, is picturesque but does not fit Shannon's description. His small pool may
have long since succumbed to agriculture, assuming it was actually in this area.

FIGURE 3.2. Wooden Mormon odometer similar to one Captain Hunt used to
measure the Southern Route.

Several emigrants comment on the decorated pottery shards found from Pavant Valley south into Parowan Valley. Welch wrote in the Fillmore area on October 14: "Several pieces of broken pottery.... far superior to any possessed by the present Indians" (Ressler 1964, 265). Sheldon Young "discovered the ruins of an ancient city [where] there was Eart[h]ine [pottery] and glass · it was 5m in extent" (October 14, 16, 1849). Two days later he "found plenty of earthan ware" as they passed through Dog Valley and near today's Cove Fort, and Lorton mentioned seeing some farther south in the Beaver River valley: "The pottery, or pieces of the same found thro[ughout] this country is artificial as can be seen from the rim & flowers painted on them & impressed figures. Naturalists attribute it to an intelligent race that once inhabeted this region" (October 23, 1849). Years later, Jayhawker Charles Mecum was still curious about them: "Have any of you ever come to any definite conclusion as to that phrenomonon we saw after we struck off beyond little Salt Lake [Parowan Valley]. there we first saw posative signs that at some period there had dwelt there a more inteligent race than now inhabited those parts" (1872).

The Fremont Indian culture in this area flourished about AD 400 to 1350 and was named for the Fremont River in south-central Utah, where the first archaeological work was done. The Fremont beginnings predate the Pueblo people, and although they lived adjacent to them, they were distinct from the Anasazi culture. The Fremont Indians became agricultural (corn, beans, squash) but still clung to hunting and gathering. They were not unified, often living in family units with a few pithouses at each "village," but also in large assemblages such as the one excavated on Five Finger Ridge in the Clear Creek Canyon about four miles west of the community of Sevier. They are known for wearing moccasins rather than sandals and for their thin-walled gray coiled pottery and detailed trapezoidal figurines and petroglyphs. David Madsen said "The key to understanding the Fremont is variation"(1989, 63). Drought is considered the probable cause of their abandonment, or absorption by encroaching cultures, but they are thought to be ancestors of a wide range of modern western Indian groups (Marwitt 1986).

Relics were of interest, but a more immediate concern was finding something to eat. The emigrants found and feasted on the numerous hare in Pahvant Valley. "Had fine Sport killing Mountain hare," said Young, and in Beaver River valley Lorton killed enough to "load my pritty Mazeppa with rabits down to the guards making its pritty white coat red with blood" (October 20). Lorton named his steed after seeing the 1846 Currier and Ives illustrated edition of Byron's epic poem *Mazeppa* that depicted a plunging white horse. In the poem, however, Mazeppa was the man; the steed had no name.

FIGURE 3.3. Beaver River near Adamsville, Utah, looking east in early summer instead of October when the Jayhawkers were here.

Meanwhile, the large, unwieldy wagon train Lewis Manly called the Sand Walking Company was beginning to fray around the edges. Welch recorded:

> The Col [Baxter] decided that we should lie in camp to keep the Sabbath and give the sick men who are improving an opportunity to rest. A number of the train manifest great uneasiness at the delay.... The meeting was then addressed in a short pittey [pithy] speech by Maj Hunt in which he urged the absolute necessity of subordination and submission to the standing authorities. The only safety said he on our long and dangerous journey is perfect unity and obedience. (October 14, in Ressler 1964, 265)

By October 20, near today's Adamsville in the Beaver River valley, Young wrote they were "210 ml [miles from] the city of the great Salt Lake."[3]

South of the Beaver River valley, the Black Mountains presented a rugged barrier to the emigrants' passage. Although the San Joaquin Company was supposed to be the first train down the Southern Route, a precocious group of wagons, the Independent Pioneer Company, was ahead of them by a few days. Previous to them, no wagon had crossed the Black Mountains. However, the year before, Porter Rockwell guided thirty-four men, the Davis family and their light wagon, plus a herd of 135 horses and mules around the west end of the Black Mountains (L. and J. Johnson 2014, 28). Hunt wanted to find Rockwell's route to eliminate the difficult mountain crossing, and thus began a disastrous detour down Beaver River and into the Escalante Desert.

Unfortunately, Hunt did not have a clear memory of the topography of the Escalante Desert in relation to Cedar and Parowan Valleys to the east, and both he and the emigrants relied on Frémont's 1848 map, which gave a totally misleading picture of the topography. As Carl I. Wheat said in 1956, the 1848 map was "the single case in which Frémont and Preuss failed to follow their standard course of showing cartographically only the geography actually seen by them, the experiment resulted in disaster." Both Lorton and Welch later make clear neither Hunt nor the emigrants realized there were *three* valleys, not just two: Parowan Valley (Little Salt Lake valley), Cedar Valley, and the westernmost valley called the Escalante Desert.

Blundering into the Escalante was pivotal in hastening the disintegration of the company and the ensuing destruction of property and untold misery of its members. On average, half of all goods and valuables of every person in the train were lost by the time they reached the coast via the Old Spanish Trail. They all suffered from near starvation and from sickness caused by bad water, and many suffered frostbite (Granger 1850, 17, 20). But far more was lost by those who attempted to reach Walker Pass by heading west instead of turning back

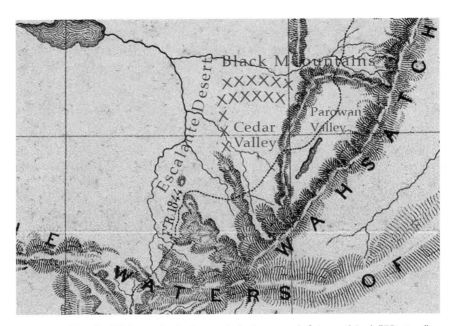

FIGURE 3.4. Detail of Frémont's 1848 map. Lake in upper left is mythical. "FR 1844" is along the Antelope Range; thumb to the right (connecting with the *X*s) are Three Finger Peaks hills. Red Hills separate Parowan and Cedar valleys. *X*s represent topography missing from Frémont's map—the western Black Mountains, and the Mud Springs Hills and Bald Hills that connect with Three Finger Peaks to enclose Cedar Valley.

from Mount Misery to follow the Old Spanish Trail. Four of the Death Valley '49ers died. Nine of the Savage-Pinney packers may have lost their lives west of Nevada's Sheep Range (though where is unclear). Thus, the aborted Escalante Desert "route" (also a rout) was dubbed "retreat valley" and "the Valley of Errors."

During this debacle, the Orson Kirk (O. K.) Smith pack train caught up to the wagons near today's Minersville, Utah, and Smith carried a sketch map of a route that purportedly cut months of time and five hundred miles off the circuitous Southern Route to the mines (Hafen and Hafen 1954, 35–37).[4] This map was a concoction of Elijah (Barney) Ward, who claimed he had traveled the route more than once. His route left the Old Spanish Trail at some point and traveled westward across today's Nevada and through Walker Pass in the southern part of California's Sierra Nevada. This scrap of paper was the proverbial straw that broke the back of the San Joaquin Company.

The so-called Walker cutoff was a trade route used by Indians from the eastern California deserts into California's San Joaquin Valley, which later became a horse-thief trail. It was named for Joseph Reddeford Walker (1798–1872), who traversed the pass in 1834 with fifty-two men, 315 horses, forty-seven cattle, and thirty dogs—the latter destined for the cooking pot (Leonard 1978, 196). The Mexicans told Zenas Leonard, who traveled with Walker, that the Indians "prefer eating domesticated horses because the act of stealing them gives their flesh a superior flavor—and it would be less trouble for them to catch wild horses, if they could thus gratify their stealing propensities" (ibid., 191). These desert Indians did not steal horses for trading, as did the Ute of Utah. They stole for survival and to prove prowess. In Panamint Valley the Jayhawkers saw "a vast heap of offal, and the bones of horses were scattered about freely, showing that the natives were accustomed to visit the great ranges of California for their regular supply of meat" (John Brier 1903, 334–35).[5]

Late in life John Colton said, "We were traveling by a map made by Frémont.... A straight western course would take us to the San Joaquin valley, which was within 300 miles of the mines. This route was marked 'unexplored' on Frémont's map" (March 9, 1915). (Due east of the San Joaquin Valley were the Mariposa mines, the southernmost mines of the gold region. They were about two hundred miles north from where the Kern River enters the San Joaquin Valley at today's Bakersfield.)

Although Walker Pass is not labeled on Frémont's 1848 "Map of Oregon and California," his route across the southern part of the Sierra Nevada over Walker Pass is shown. "Walker's Pass" is, however, labeled on Mitchell's 1846 pocket map, which was "of real importance, highly popular, and doubtless published in a large edition."[6] It is likely some of the emigrants had a copy.

On Frémont's 1848 map, a fictitious mountain range stretches westward from the "Wahatch Mountains" almost to the Sierra Nevada. It supposedly marked the southern boundary of the Great Basin, but Frémont added this cautionary note: "*These mountains are not explored, being only seen from elevated points on the northern exploring line*" (emphasis added). Barney Ward's trail purportedly followed the southern margin of Frémont's mythical east-west mountain range.

Addison Pratt noted on October 22 that Ward's map and testimony showed a trail that "was reported to be a much nearer route than that we were going and no dry deserts to pass through, but grass and water all the way" (1990, 383). Welch also described the purported trail: "Capt. Smith who is their leader has furnished me with a diagram according to which there is wood water and grass at intervals of twelve or fifteen miles. This runs due west and it cannot be more than 400 miles to the Valley of San Joaquim. Mr. Ward of Salt Lake City has been through (twice) and has assured many that it can be traveled with wagons" (October 26, in Ressler 1964, 268–69). (Welch consistently spelled the company's name *San Joaquim*.)

Lorton reasoned that "wherever there is a chain of high mountains there

FIGURE 3.5. Side of Jefferson Hunt monument built about 1958 with colorful rocks surrounding the plaques. Here, one hundred San Joaquin Company wagons turned west from Hunt's track down the Old Spanish Trail. (The 113 number is not supported by data).

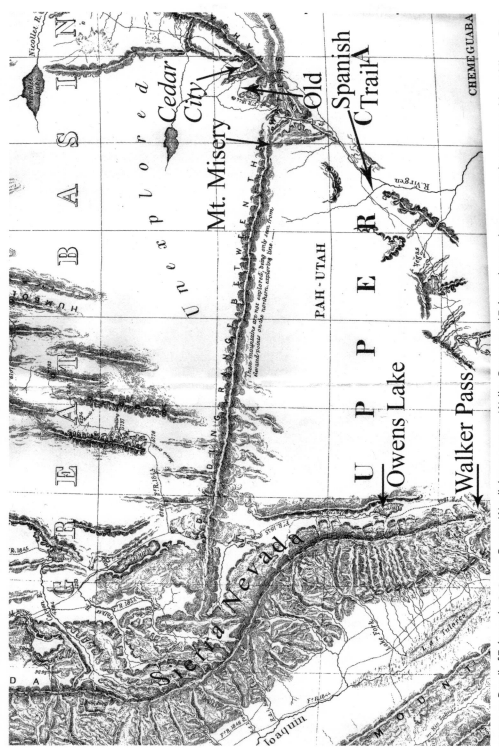

FIGURE 3.6. Detail of Frémont's 1848 map. Barney Ward's bogus route to Walker Pass crossed Frémont's nonexistent mountain range near Mount Misery. But Frémont cautioned, "These mountains are not explored, being only seen from elevated points on the northern exploring line," a warning the emigrants ignored.

must be snow, and wherever there is snow there must be streams emitting therefrom" (January 20, 1850). Lorton and others ignored the fact that Frémont's "mountains are unexplored." Thus, the very tempting shortcut was widely discussed and eagerly received—and Frémont's warning was readily pushed aside.

Unfortunately, these emigrants did not have the advantage of Pegleg Smith's sage advise. Smith knew Barney Ward and said he was a liar and that his stories were false (Stuart in Hafen 1969, 7:347; Webster 1917, 72).

On October 23 Young "lay in camp all Day" in the Escalante Desert, waiting for Hunt to find the trail. This was his last entry until November 9 at Mount Misery. The same day Welch recorded:

> The guide [Captain Hunt] returned late last night and reported that he had found no water for twelve miles ahead. In consequence this morning all was confusion and dismay.... Men were gathered in groups around the camp discussing the best means for extricating themselves from the present difficulties in a state of high excitement some blaming the guide others the Council.... Advise the divisions to drive their cattle back to Beaver Creek.... The cattle were in a state of extreme suffering when they arrived but are now feeding quietly on the bank. There are multitudes on the creek without their tents who will be obliged to camp out. (October 23, 1849, in Ressler 1964, 267)

By October 26 the train retreated from the "Valley of Errors" back into Beaver River valley and started across the Black Mountains, "the devils own pass," following the Independent Pioneer Company's trail up California Hollow, south of today's Greenville, Utah.[7] Near the summit camp in the virgin juniper forest (about 6,940 feet in elevation), Welch commented that "the Jay Hawkers are dancing in grotesque attitude to the twang of an old fiddle. All seem cheerful from the certainty that we are again on the right road" (October 26, 1849, ibid., 268).

In Parowan Valley, east of Little Salt Lake, the train intersected the Old Spanish Trail. Welch said the Barney Ward cutoff "turns west from the old spanish trail seventy miles from L S L [Little Salt Lake]" (ibid.).

From the copious springs at today's Enoch, the pack trail crossed soggy Cedar Valley to Iron Springs and thence around the northern toe of Antelope Range. The wagons had to detour south to today's Cedar City to circumvent the boggy part of Cedar Valley. Near Cedar City, the train fragmented.

> Capt Doty commander of the Jay Hawkers rolled out with eight wagons of his division & Capt [Cephas] Arms is intending to leave the San Joaquim

Company with his entire division [the 1st, or Buckskins] to morrow. The Col [Baxter] as his first obligation lie[s] with the Rovers who are members of the first division has resigned his office [as captain of the wagon train]. Considerable excitement is exhibited in camp. The reasons which Capt. Arms gives for leaving is the exceedingly slow progress which so large a company makes and the unwillingness of his division to incur the risk of traveling through the deserts with men who have weak teams and are short of provisions. (Welch, October 30, ibid., 271)

The seven divisions, however, kept their internal cohesiveness for another few days and some all the way to the coast.[8] Welch's Buckskins were now a day ahead of Lorton's division, and the Jayhawkers were apparently a day ahead of Welch; the packers were a day or more ahead of the Jayhawkers.

Many of the gold rushers labored under the delusion that each day spent on the trail was one day less they could pick up gold nuggets along the Sierra Nevada streams. After losing faith in Captain Hunt, they succumbed to the temptation of the purported shorter route to the goldfields, and the words "These mountains are not explored" were easily ignored. "The disposition of the human mind to accept the marvelous without question" infected the whole train (*Inyo Independent*, October 4, 1884).

Lewis Manly summed up the fateful decision to secede from their guide and venture into the unexplored wilderness:

Cut-off fever began to rage in camp again.... His [Barney Ward's] plan grew so much in favor that when the place was reached a hundred wagons turned out into the Smith trail, leaving Capt. Hunt only the seven Mormon wagons bound for San Bernardino. Hunt stood at the forks of the road as the wagons went by and said to them;—"Good-bye, friends. I cannot, according to my agreement go with you, for I was hired for this road, and no other was mentioned. I am in duty bound to go even if only one wagon decides to go."...He wished them luck and the two trains parted company. (1894, 327–28)

Lorenzo Dow Stephens vividly recalled the moment: "When we started on [this route] Captain Hunt called out to us 'boys if you undertake that route you will go to hell,'" a prophetic parting volley (1916, 17, 19).

The turnoff from the Old Spanish Trail is now commemorated with the Jefferson Hunt monument about seven miles northeast of today's Enterprise, Utah. Addison Pratt wrote, "Old Spanish Trail continues south. Here the waggons began to file off onto the cutoff [to Walker Pass], till they left but 7 to go the old trail" (1990, 386).

The seceders entered Shoal Creek Canyon three miles west of today's Enterprise, where the grade was gentle, with plenty of water, firewood, and forage for hundreds of livestock—just as Ward described the route to O. K. Smith.

About nine miles up the valley Lorton found a distinctive cave he called "Fireplace Rock." That and his sketch labeled "Cave at Cave Canyon" provide irrefutable proof of the route up Shoal Creek, through today's Terry Ranch, and into Nephi Draw (later named for southern Utah pioneer Nephi Johnson).

At the head of Nephi Draw, the wagons emerged onto the flat white rise they were sure was the eastern end of Frémont's east-west mountain range. The wagons followed the packers' trail and turned southward toward today's Pine Park Bench as "a light snow had just melted rendering the ground so soft that the cattle sunk hoof deep and our progress was consequently slow" (Welch, November 5, in Ressler 1964, 274).

Ward's map made clear they were to cross to the south side of the range and follow it westward. Packer George Cannon clarifies their route: "[It] struck West thro' the unexplored region laid down by Fremont on his map, & south of [a] range of mountains described by him as running East & West.... [We] turned our course Southward over the dividing ridge of the Great Basin...[and] came

FIGURE 3.7. Fireplace Rock on Terry Ranch in upper Shoal Creek valley below the mouth of Nephi draw, 2006. "A cave resembling an old fashioned fireplace & chimny but no hole to emit smoke" (Lorton, November 4, 1849).

to a large Cañon [Headwaters Wash] running W. S. W. down which we went until we came to the Devil's offset" (1999, 23, 38).

Two of the wagon scouts followed tracks of the O. K. Smith and Mormon packers, but they brought back discouraging news. "They had discovered no water and the hills had become so precipitous that it was necessary to travel in the canons [canyons]. One of these which lay on our course they had explored but it led to a narrow gorge [Cannon's Devil's offset] so filled with huge rock as to make a passage with wagons impracticable" (Welch, November 5, in Ressler 1964, 274).

The splintered San Joaquin Company was now in complete disarray. The westward progress of the wagons was blocked by the precipitous walls of White Hollow, Pine Park Canyon, and Headwaters Wash. They could no longer follow the packers! The one hundred wagons spread southward for two miles along the elevated, sparsely wooded Pine Park Bench the emigrants named Mount Misery.[9]

CHAPTER 4

MOUNT MISERY

Great many discouraged & know not what to do.
—William B. Lorton, November 7, 1849

THE REQUISITE TO UNSCRAMBLING the route of the Death Valley Jayhawkers through Nevada is to first correctly locate Mount Misery in southwestern Utah. With information from firsthand accounts—those of Lorton, Welch, Manly, Cannon, Bigler, Nusbaumer, Brier, and several Jayhawkers, plus at least twenty trips to investigate the area on foot—LeRoy and I conclude the location of Mount Misery is the two-mile belt of rolling hills now called Pine Park Bench. It is about twenty road miles west of today's Enterprise, Utah, and two miles east of the Nevada border just after crossing out of the Great Basin. It extends south from a hill called Roundup Flat to the brink of Pine Park Canyon and White Hollow and sits on an unusual geological formation, part of an extensive layer of tuff (at least a thousand feet thick) ejected from the Pine Park Caldera about twelve million years ago. At the south end of the bench, the precipitous walls of White Hollow display spectacular hoodoos.

Lorton: "For 2 miles strung along up the hill tops appear all the old comp[any]" (November 5, 1849). Welch added more information about the high benchland that became known as Mount Misery: "Word has come tonight that [there] is a fine stream running in a deep grassy valley or gorge two miles south of us [Pine Park Creek in White Hollow]" (November 5, in Ressler 1964, 274). Some of the '49ers called it Poverty Point or Deep Diggings and recalled it as the place where they had a miserable time with cold temperatures and difficult-to-reach water sources (Stephens 1881, 1908; James Brier 1886).

Researchers before LeRoy Johnson who looked into the location of Mount Misery included Charles Kelly, Dale Morgan, Margaret Long, George Koenig, and Todd Berens.[1] Kelly located the initials of Henry W. Bigler, one of the packers, inscribed on a wall of volcanic ash in Headwaters Wash (1939, 6–8, 41, 43). We, however, have had the advantage in our research of using the invaluable first-hand information from the journals of William Lorton and Adonijah Welch that were not accessible to most of the others.

The packers crossed the rim of the Great Basin and found huge boulders blocking their progress in upper Headwaters Wash. They retreated up the wash, circled Roundup Flat, and descended today's Clay Canyon into Headwaters Wash below the boulder blockage. However, the packer route was impossible for wagons to follow. Thus, the one hundred wagons threaded themselves along the two miles of high, rolling benchland. Nusbaumer recalled on the third day after leaving the Old Spanish Trail, "We came across nearly all the other parties camping on knolls without water and without food for the cattle, on the edge of a deep and rocky cañon which we were unable to cross. It took us an hour to carry water up the hill and the cattle were in danger of breaking their necks when driven down the trail" (1967, 34).

There is no water on Pine Park Bench (Mount Misery), but about two miles south of Roundup Flat, scouts followed an Indian trail into White Hollow and found today's Pine Park Spring and its stream with forage nearby. Today's U.S. Forest Service trail is a slight realignment of the Indian trail. Lorton wrote in his journal: "At last we reach the bottom with ponies & cattle. Here we find a creek where the cattle drink. Several cattle got hurt by falling from the precipices.... Greatest hill I ever see. No more news till eve'g. Explorers come in & report that

FIGURE 4.1. Rolling hills along Pine Park Bench (Mount Misery) near head of Clay Canyon where wagons parked to search for a wagon route when they could no longer follow the packers. West of Enterprise, Utah, near Nevada border. Most of the San Joaquin Company returned to the Old Spanish Trail, but the Death Valley '49ers found a route around the labyrinth of canyons and continued west, and some men became packers.

it is impossible to pass with wagons. Great many discouraged & know not what to do. Some women [take the news] better than the men" (November 5–7). James Brier: "We had to bring our water up that fearful slope [so] we named the place 'Deep Diggings'" (1886). John Brier: "The only man who could descend to the stream was a Canadian voyageur, and those who drank of its water were compelled to pay at the rate of one dollar per bucket" (1903, 328). Charles Mecum: "All will remember the deep canyon where the little Frenchman carried up so much water" (1872). Dow Stephens: "A small stream flowed at the bottom, and by using ropes and buckets we could get enough for camp use, using it sparingly" (1916, 19).[2]

The depth of the canyon from brink to stream in White Hollow is 560 feet, although Lorton felt it was more: "I decended this gulf or ravine, I suppose 2,000 ft, very steep on one side" (November 5).

During the forlorn days spent at Mount Misery, numerous men scouted for a wagon route westward. The inscription "OSbORN '49," in a side canyon or gulch of Headwaters Wash, identifies a route scouted by William Thomas Osborn, one of the Bug Smashers who later made it over Walker Pass to the gold mines.[3]

Manly wrote, "They named this place Mt. Misery. While camped here a lone and seemingly friendless man died and was buried...think he was from Kentucky" (1894, 329). Lorton: "Poor Naylor died. Quarel over his body for his property. I reprove them.... Burry him on the mountain. Several wagons left & [a] great many splicing" (November 5). Lorton refers to "Nailer" earlier on the trail and seems to have felt some responsibility for Naylor's possessions after his death. Thanks to Lorton, the name of this solitary black man now has its rightful place in western history.[4]

On November 10, Cephas Arms stated what was, by then, the emigrants' general opinion of Barney Ward and his map: "We are satisfied that it is all a humbug."

The wayfarers had four options: abandon their wagons on Mount Misery and continue westward on foot or as packers, return to the division point (today's Jefferson Hunt monument) and follow the Hunt train down the Old Spanish Trail, return to Salt Lake City, or find a wagon route around the canyon complex and continue westward.

As scout after scout returned without finding a wagon route around the chasms of Pine Park Canyon and Headwaters Wash, about seventy wagon masters gave up the "folly of exploring with ox-teams" (Arms, November 10). Manly wrote that a Mr. Rynierson said, "My family is near and dear to me.... [I]f we were to remain here and be caught in a severe storm we should all probably perish. I, for one, feel in duty bound to seek a safer way than this. I shall hitch up my oxen and return at once to the old trail" (1894, 112). Over the next couple of days, more than two-thirds of the San Joaquin Company retreated to the Old Spanish Trail, some arriving November 13, "having lost ten days travel and provisions" (Arms, November 10).

Other men dismantled their wagons and made packsaddles for horses or mules and followed the O. K. Smith and Flake packers who were a few days ahead of them. William Lorton was one of these men. "Our teem gave out.... Conclude to pack & make preparations.... All ready & start down the Big hill with the pack on Mazeppa's back & calculated to meet the five New York boys 2 miles down" (November 10 or 11).[5] There were assorted other men, including eleven later called the Pinney-Savage party, who also became packers.[6] At the base of the Sheep Range in southern Nevada, the eleven continued west on foot, while all the other packers found their way back to the Old Spanish Trail by various routes. Both Pinney and Savage made it to the goldfields, but the fate of the remaining nine who started with them is a mystery, although rumors of skeletons and remains have been recounted.[7]

Scouts from the remaining wagons (over a quarter of the original San Joaquin Company) found a route north around the tortuous maze of canyons. Manly used the number "26 wagons" in a notation on a map he sent to John Colton in 1890, and he also used the same number in his serialized account, "From Vermont to

FIGURE 4.2. Hoodoos at White Hollow on the way to Pine Park Campground below cliffs of Mount Misery eroded in the twelve-million-year-old volcanic Tuff of Honeycomb Rock.

California." But he used "twenty-seven wagons" in *Death Valley in '49*, and secondary references have used this latter number.[8]

Cephas Arms recorded on November 10 that Jayhawker Aaron Larkin "came to our camp to inform us that they had found a valley which they thought would lead out of the mountains, and they are going to move forward."

At first, the Bennett family wagons also turned back toward the Old Spanish Trail. According to Manly:

Shortly after they [Ryniersons and others] had started, a delay was caused by the death of a man who was taking the trip for his health [Mr. Naylor]. All of our party had not got back from hunting the pass. Mr. Bennett and some others were slow in starting and when we had got only a short distance from camp Mr. Bennett broke a wagon axle and had to return to get one from the wagon left by the man that died. Before we got the wagon fixed news came from those who had returned to camp that the pass was found and that they would go on ahead. So they hitched and turned nearly due north. When Mr. Bennett's wagon was mended those that had waited, as well as himself, followed on the new route. Our little Green river party was now divided and only Rogers and myself were together [with the two Bennett wagons]. (December 1887)

CHAPTER 5

MOUNT MISERY TO DEATH VALLEY

A Dismal looking country…a Dubious looking country.
—Sheldon Young, November 16–17, 1849

You all remember the time we was camped on Poverty Point,
where the most of the train turned back to the Old Spanish trail.
better for us had we all turned back.
The Jay hawkers counseled together and determined
to go ahead or leave their bones on the Plains.
—Lorenzo Dow Stephens, 1881

JAMES BRIER SPOKE of Mount Misery and their departure: "Our progress was blocked by high mountains and terrible gorges; and we finally decided to turn back. After some prospecting, however, I found open country to the west. My company—I was in with a lot of free-hearted, Mississippi boys—voted to go that way, and we started, with the Jayhawkers following" (1886). The Mississippi boys were part of the group I collectively call the Bug Smashers, the name Sheldon Young used.

From Mount Misery, about one hundred emigrants doggedly rolled north-westward, venturing into a region where no wagon—or white man—had gone before. They did, indeed, fulfill Hunt's prophetic remark: "If you undertake that route you will go to hell" (Stephens 1916, 17, 19).

The Death Valley '49ers were composed of very loosely connected groups—four families, the Platte Valley Jayhawkers, part of the Bug Smashers division of the San Joaquin Company, hired drovers, men who spoke German, and assorted others such as Sheldon Young's mess.

The Bennett and Arcan families consistently traveled together (four wagons), and others, such as the Earhart brothers and son, joined them.[1] At some point the Nusbaumer mess became part of this small train, as did Anton Schlögel's wagon,

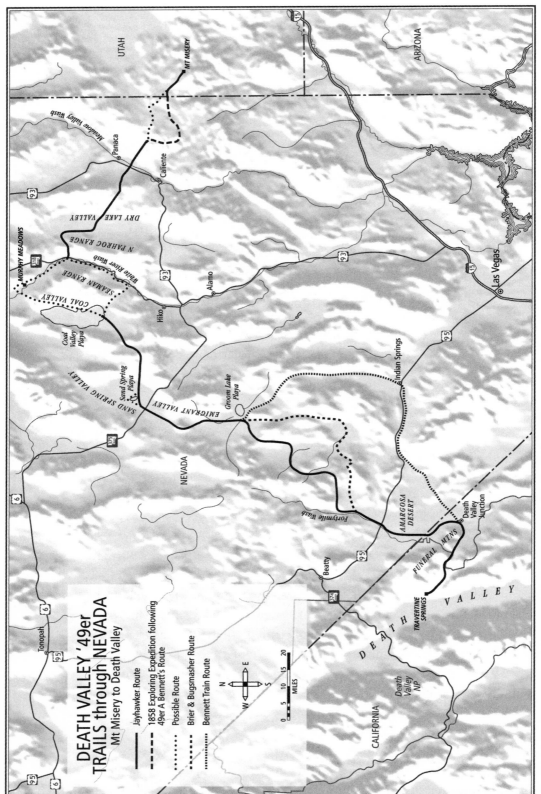

FIGURE 5.1. *Death Valley '49er Trails Through Nevada: Mount Misery to Death Valley*

and the Wade family followed in its wake, often a day behind. Henceforth this group is referred to as the Bennett Train. Upon entering Death Valley, most of them turned south and thus became the southern contingent of Death Valley '49ers.

Over time these '49ers congealed roughly into three groups: the Jayhawkers (and men who traveled near them),[2] the Bug Smashers (and Brier family, who traveled with or near them), and the Bennett Train. The Jayhawkers were described in the introduction, and this is a good time to meet others who were part of the northern contingent (those who turned north in Death Valley to find a way over the mountains before them).

The Brier Family

James Welch Brier, his wife, Juliet, and their three young sons left home in Maquoketa, Iowa (thirty miles south of Dubuque), on April 19, 1849, heading for California so James could preach the word of a Protestant God to the swarm of gold seekers and new settlers in Catholic Alta California. In early 1849 the Iowa Methodist Board solicited volunteers to organize churches in California, and James rose to the call. In a letter home, Juliet recalled, "James thinks it his duty to go on account of the great emigration and the small supply of ministerial labor. I was greatly opposed to it at first but am becoming reconciled and even anxious to go on" (May 19, 1849, in Leadingham 1963–64, 61).

The Brier family became Death Valley Jayhawkers by default. After the breakup of the San Joaquin Company at Mount Misery, James decided to continue toward Walker Pass. The Jayhawkers did not want to travel with women and children, so the Brier family fell in with the Bug Smashers. However, after leaving Death Valley, the Bug Smashers separated from the Briers in Panamint Valley, just west of Death Valley. From then on, an ebb and flow of different men, including some of the Jayhawkers, traveled with the Briers, and the association became closer as they proceeded.

Juliet Wells Brier, the Death Valley '49er woman about whom most is known, was born of hardy New England stock on a farm near Manchester, Vermont, on April 8, 1814 (James Brier 1893; Juliet Brier 1904; Leadingham 1963–64, 13; U.S. Census 1900 for California). Thus, she was thirty-five while crossing the plains and western deserts.

Juliet, known to her friends and family as Julia, was schooled at Bennington Academy (later Seminary) in Vermont, where she studied the "English Course"—spelling, reading, writing, geography, history, and mental arithmetic (MacBrair-Koller 1988).

Juliet was small, never weighing more than 115 pounds, and of dark complexion, with hazel-brown eyes and red highlights in her auburn hair. She possessed a spirit that bloomed and grew in the heat of adversity; those who traveled with her through the Great American Desert remembered her as hardworking and generous. "All agreed that she was by far the best man of the party. She was the one who put the packs on the oxen in the morning. She it was who took them off at night, built the fires, cooked the food, helped the children, and did all sorts of work.... [I]t seemed almost impossible that one little woman could do so much. It was entirely to her untiring devotion that her husband and children lived" (Manly 1894, 342). Her faith was always a comfort to her, and her keen mind and memory brought forth scripture at will (Hilda Brier, Juliet's granddaughter, in Leadingham 1963–64, 15).[3]

Juliet moved from Vermont with her pioneering family to Mottville in southwestern Michigan in the 1830s, where she probably taught school while living at home. She married the Reverend James Brier, a Methodist circuit rider, on September 23, 1838.[4] They moved often during the next ten years, a pattern that continued in California. Juliet wrote her brother that "California fever... [is] raging all over the circuit" before she had any inkling she was about to become part of it.

The Briers joined the San Francisco Company, where "Rev. Jas. W. Brier" was listed as the company chaplain, and an addendum to the roster listed eleven women and twenty-three children, including "Mrs. Brier and three children."

Juliet explained why they went to California, the difficulties of just getting through Iowa, and how the family was thriving. It is not only a personal account but describes what many other pioneer families experienced as well:

After six weeks of toil in traveling over almost impassable roads with oxen, camping out exposed to heat and cold, we have with hundreds reached the bluffs on the Missouri.... We have 4 yoke of cattle one of cows & a cow that follows the wagon, plenty of flour, sea-bread, bacon, dried meat, rice, sugar, tea coffee & dried fruit and beans....

I should like much to give you a description of our journey thus far. I have seen and heard more that is new in the six weeks [past] than in any other whole year of my existence. You may imagine 20, 30, 40 and sometimes 50 waggons with 4 or 5 yoke of cattle moving on slow over broad prairies deep ravines timbered portions, low wet river bottoms and every variety you can mention.

These waggons mostly contain three men and provisions for a year.... [W]e have been under the necessity of leveling roads, building bridges, letting our waggons down steep pitches by ropes, doubling teams to draw up and out of mudholes and sloughs, many times drawing waggons by

hand over temporary bridges, occasional breakdowns, upsetting of waggons, mashing our tin dishes, mixing our groceries, etc, etc. But notwithstanding all these I feel as much at home as I ever did in my life—the
children are all fat and healthy and very happy. (Leadingham 1963–64, 61)

Juliet's letter alternates between her personal misgivings and the assurances
she wants to give her family. In this vein she told them, "This is a new way of
living and not altogether unpleasant.... You must not be troubled about us but
pray for us, we know not what may befall us, but God is able to protect us." After
promising a lock of hair from each of her sons, she reminded them, "Do not
forget to pray for us!"

The San Francisco Company crossed the Missouri River at Traders Point,
about ten miles south of Council Bluffs, with thirty-five wagons, 231 oxen, seventeen cows, seventeen horses, four mules, 149 firearms, and twelve stoves,
along with tons of breadstuff, meat, groceries, gunpowder, lead, and medicine
(*Dubuque Tribune*, August 10, 1849). Juliet wrote to her family from Council Bluffs
on May 29, before she headed into Indian country. That same day the Asahel
Bennett family left Council Bluffs in the Badger Company, and the Knox County
Company, including the future Jayhawkers, had left just a week before, on May
22 (Rasmussen 1994, 1:67–68).[5]

On the Sweetwater, west of Independence Rock, the Briers apparently separated from their wagon train, possibly over a typical point of disagreement—to
travel, or not to travel, on the Sabbath. Sarah Royce wrote in her diary three days
after arriving at Independence Rock:

On Sunday, the 29th of July, we determined to remain in camp and rest till
the next day. One family of our fellow travelers, Mr. B——— and his wife
and three little boys, did the same. We enjoyed a quiet rest, held a social
meeting for prayer, reading and singing, and the next morning resumed
our journey, much refreshed. From this time till we reached Salt Lake we
had no earthly company or protection except that mutually afforded and
enjoyed by two men, two women and four children, the oldest not more
than eight, and the youngest not yet three.... We passed the company
we had been traveling with, kept in advance of them, notwithstanding
Sunday rests, and arrived in Salt Lake valley [August 19] the day before
they did. (Royce 1939, 26)

The Brier family arrived eleven days after the Jayhawkers. Traveling without the
protection and support offered by a large wagon train may have caused Julia anxiety and misgivings, but it prepared her for a far more lonely venture yet to come.

James Welsh Brier was born near Dayton, Ohio, on October 14, 1814. He graduated from Wabash College in Crawfordsville, Indiana (northwest of Indianapolis), and from Lake Theological Seminary in Ohio.[6] He became a circuit preacher for small congregations in hamlets and rural areas, and Jacob Stover recalls Brier preaching "in Iowa City and Pleasant Valley, a tall, slim man" (Stover 1937, in Hafen and Hafen 1954, 277). Years later, he was described as "a tall, raw-boned man of intelligent face and natural eloquence whose countenance bears many a graving of hardship, but who carries his seventy-three years with wonderful lightness" (James Brier 1886). The Great Register of Lodi gave his description as five feet eleven and a half inches, fair complexion, and blue eyes (Charles Brier 1956, 46). Manly said, "He was a minister, a very intelligent cultured gentleman" (1884). James was forceful, but he lacked a certain empathy toward his wife's well-bred sensitivities when he insisted on traveling in company with other men that included no women for Julia's moral and physical fellowship during three dreadful months after they left Mount Misery until they arrived at the Rancho San Francisco in Southern California.

The Briers' three sons were Christopher Columbus, born September 11, 1841, in Indiana; John Wells, who followed his father's career as a minister, born May 28, 1843, in Michigan; and Kirke White, born May 5, 1845, in Indiana.[7] Thus, the boys were ages nine, six, and four when they traveled through the Nevada-California deserts.

James Brier suffered severely from diabetes after leaving Panamint Valley. Before that time, he carried his children, herded his newly procured oxen, and walked extra miles to scout the route. Jayhawker accounts treat Brier poorly, partly because he was a vocal proponent of this disastrous "cutoff," and human nature looks for someone to blame. They also did not realize that diabetes, not laziness, made him an invalid. During the last days of their ordeal, Juliet had to lift her husband to his feet in the morning so he could stumble on with the aid of two canes. He lost more than one hundred pounds—wasting away to seventy-five pounds (James Brier 1890). It is a wonder he survived the final 150 miles of the trek on foot.

Colton considered the Brier family part of the Jayhawkers when he started the Jayhawker reunions in 1872. He tried hard to find their address to invite them to attend, and after he succeeded James Brier wrote long letters for the reunions.[8] Juliet not only wrote letters but also hosted the 1903, 1908, and 1911 reunions in Lodi, California, where she lived with her son John after James's death in 1898. She also received a visit in Santa Cruz from John Colton and Dow Stephens during the 1913 reunion shortly before she died on May 26, 1913 (*San Francisco Call*, February 9, 1913).

Meet the Bug Smashers

Turning north in Death Valley was a group of men from Mississippi, Georgia, and other states. Sheldon Young called them the Bug Smashers, but they were also referred to as the Georgia-Mississippi boys (E. I. Edwards), the Georgians (Dow Stephens), the Martin-Townes group, or the Coker party (in Manly 1894). For convenience, I lump together the Mississippi boys, the Georgia men, Coker from New York, Martin from Missouri, and others under the title "Bug Smashers," since many apparently were in the Bug Smasher division of the San Joaquin Company (Stephens 1916, 16).

Addison Pratt, traveling with other Mormons in a wagon with an odometer, was also in the Bug Smasher division. He mentions three men who later trudged through Death Valley—Mastin, Townes, and Martin:

> 12th Oct…I had no acquaintance among the gold diggers in particular, but Blackwell found a cousin by the name of Mastin, among a company of wagons from Mississippi, and as ours had to be put in a company we chose that one. Town [Townes] was the captain's name of this company. We found them to be a kind hearted, obliging company of men to travel with, and we verry often messed [ate] with some of them, especially with Mastin's wagon, and one other by the name of Martin. (Pratt 1990, 381; see also Hoshide 1996)

Lewis Manly mentions a James (Jim) Martin in a reunion letter. He remembered him as being a captain of one of the groups, and "I remember James Woods & Nat Ward—Woods worked a claim along side of me in 1850 & told me all his troubles & hardships after leaving the wagons—Ward died in San Jose with the cholera I think in 1852 or 1853" (Manly 1888). When Manly hiked north in Death Valley to find out what the northern contingent was planning, he found "the Martin party was already in marching order and…gave all their oxen they had left to Mr. Brier" (1894, 337).

The Bug Smashers were the second-largest group in the northern contingent, but their cohesion did not equal that of the Jayhawkers, and the names of some may still be missing. Several of them had been members of the California Exploring and Mining Company, organized in Pontotoc, Mississippi, that left Independence, Missouri, on May 9, 1849. Paschal Townes, for whom Towne Pass in the Panamint Range was named (but misspelled), traveled with Jackson Thomason from Independence to Salt Lake City. Thomason mentions Townes several times in his diary (1978, 23).

The California Exploring and Mining Company arrived at Salt Lake City on August 2 (six days before the Jayhawkers) and the next day "dissolved by a unanimous vote." Several of the "Mississippi boys" lingered in the city (Townes, the Turners, Osborn, and at least one man named Carr), then they joined the Bug Smasher division in the San Joaquin Company that took the Southern Route to the goldfields. E. I. Edwards says the Mississippi boys traveled with some men from Georgia (1940, 32). Dow Stephens called them Georgians and said there were "about fifteen of them" (1916, 20), and the *San Francisco Chronicle* reported the number was "about twenty," probably from John Colton (February 15, 1903; see also Wheat 1939a, 113). Edward Coker, a New Yorker, wrote Manly that there were twenty-one in the party with whom he traveled through Death Valley, but he gives only seven other names (Manly 1894, 373, 375). Some of the men from southern states brought their servants (slaves?) with them. The current list is in Appendix B.

Others Who Became Death Valley Jayhawkers

Part of Louis Nusbaumer's mess—Fish, Isham, and Gould—were included in the later Death Valley Jayhawker reunion lists. A Frenchman (whose name no one could remember) was also regularly included in the lists.[9]

Sheldon Young and his mess (Wolfgang Tauber, Frederick Gritzner, and possibly William McGinnis) may have been in the Jayhawker division of the San Joaquin Company, although they "made up our minds to stay through the winter" in Salt Lake City before being convinced to take the Southern Route (Young, July 23, September 25, 1849).

Lorenzo Dow Stephens, who arrived in Salt Lake City on July 25, two weeks earlier than the Knox County Company, may have been in the Jayhawker division on the Southern Route (Emigrant Rosters 1849, July 25, 1849).[10] We do not know who was with him or if he met Jayhawkers in Salt Lake City and joined them there. He came from Cedar Township (Abingdon), Illinois, only eleven miles south of Galesburg, and may have previously known some of the Knoxville and Galesburg men (*Annals of Knox County* 1980, 57).

As historian Carl I. Wheat said, "The [Death Valley] Jayhawker list was itself somewhat elastic, varying from year to year under John B. Colton's enthusiastic reunion promotion" (1939a, 103). The most common number was thirty-six. Colton wrote T. S. Palmer:

> By a Sudden inspiration, two names came to my mind, since the meeting, two names none of us could remember. they joined us at Salt Lake. we Knew them as well as ourselves, but not coming from the Same locality in Ill[inoi]s the names passed out of my minds—they are as follows— [Sheldon] Young & Woolf[gang Tauber]—both from Joliet Ills. I have

Young in the list as "man from Oscaloosa Iowa name not remembered, died in California."[11] this is all confirmed by Stevens of Cali—the [*illegible*] thing is that we should have forgotten Woolf—he was a good man, and held out to the last—doing a great portion of Exploring for water when life was not worth living—he & Young were in same Cart with Gretzinger [Frederick Gritzner] of Oscaloosa Iowa. (March 21, 1894)[12]

Colton's numbers changed, Manly's list is not as long as Colton's, and researchers from Ellenbecker on have included different names. Some men joined the northern contingent as late as Death Valley (part of Nusbaumer's mess), and the Bennett-Arcan teamsters caught up in Searles Valley (two valleys west of Death Valley).

Other men who traveled in proximity to the Jayhawkers suffered the same experiences, but we know less about them. These include Goller and Graff, two Germans; Bennett and Arcan's teamsters; and possibly others not yet verified, such as Anderson Benson and William McGinnis.

Westward from Mount Misery

James Brier (1853) provided the following information on the trek beyond Mount Misery: "The company to which I was attached...[discovered] an open country to the west...[of Mount Misery, and we] resolved on a western course. Twenty miles brought us to the first Muddy [a stream]. In traveling this twenty miles, we found no serious obstacles, excepting a cedar forest, through which we cut a road.... [Then] the country over which we passed was a succession of valleys, separated by low dividing ridges." Today's Clover Creek is about thirteen miles from Mount Misery, and Meadow Valley is about thirty-five air miles, so already the data for this part of the trek is confusing.

Manly provides much of the detail on this segment of the trail, so it is appropriate to have him tell the story. Some comes from the December 1887 issue of "From Vermont to California," his serialized account in the *Santa Clara Valley*, a horticulture and viticulture magazine, and some is from his 1894 book, *Death Valley in '49.*

[At Mount Misery,] when Mr. Bennett's wagon was mended those that had waited, as well as himself, followed on the new route.... A meeting was called so as to travel systematically and after a good deal of talk, it was decided to leave the families out, as some said they would not agree to help them if the new route should prove to be a hard one. I think Jim Martin was chosen captain of the organized part. He contended that we

FIGURE 5.2. Cedar Range east of Panaca, Nevada. View west over the "cedar forest, through which we cut a road" to the Chief Range west of Panaca (James Brier).

would have to travel some distance north before turning west. The trail of Captain Smith [the packers] was now some ways to the south, so we never saw it after we left Mount Misery.... Under Martin's orders we drove north three days or more on very high ground....

I soon became satisfied that going north did not take us in the right direction. I told them so but they persisted in their course. Bennett and some others concluded to take my advise and turn west. The first night Martin & Company overtook us.[13]

Thus far the country had been well watered and furnished plenty of grass, and most of them [the single men] talked and believed that this kind of rolling country would last all the way through. The men at leisure scattered around over the hills on each side of the route taken by the train, and in advance of it, hunting camping places and making a regular picnic of it. There were no hardships, and one man had a fiddle which he tuned up evenings and gave plenty of fine music. Joy and happiness seemed the rule, and all of the train were certainly having a good time of it. (1894, 331)[14]

Shortly after leaving Mount Misery, Manly described a meeting where "those who were rejected [by the single men] were Rev. J. W. Brier and his family, J. B. Arcane and family, and Mr. A. Bennett and family. Mr. Brier would not stay put out, but forced himself in, and said he was going with the rest, and so he did. But the other families remained behind" (1894, 114).

Manly's various accounts provide different bits of information, and some of it varies enough to confuse who was where at a particular time. But he provides helpful details about their trip from Mount Misery to the Chief Range, west of Meadow Valley (southwest of today's Panaca, Nevada): "We now crossed a low range [Cedar Range] and a small creek running south, and here were also some springs. Some corn had been grown here by the Indians." (This could be Clover Valley or Meadow Valley.) He also mentions "pillars of sand stone, fifteen feet high and very slim were round about in several places and looked strange enough" (1894, 113–15).[15] Such hoodoos are found along a route from Meadow Valley west to Bennett Springs on the slope of the Chief Range, although the most spectacular formations are north of Panaca in Cathedral Gorge State Park, about five miles north of their route.

Manly described a foreboding change in the landscape as they traveled through Nevada's Basin and Range Province:

Gradually there came a change as the wagon wheels rolled westward. The valleys seemed to have no streams in them, and the mountain ranges grew more and more broken, and in the lower ground a dry lake could

FIGURE 5.3. Formations south of Panaca and west of Highway 93 along a wash heading toward Bennett Springs resemble Manly's "pillars of sand stone, fifteen feet high" (1894, 115).

be found, and water and grass grew scarce—so much so that both men and oxen suffered. These dry lake beds deceived them many times. They seemed as if containing plenty of water, and off the men would go to explore. They usually found the distance to them about three times as far as they at first supposed, and when at last they reached them they found no water, but a dry, shining bed, smooth as glass, but just clay, hard as a rock.... Nothing grew in the shape of vegetables [*sic*] or plants except a small, stunted, bitter brush.

Away to the west and north there was much broken country, the mountain ranges higher and rougher and more barren, and from almost every sightly elevation there appeared one or more of these dry lake beds. One night after about three days of travel the whole of the train of twenty seven wagons was camped along the bank of one of these lakes, this one with a very little water in it not more than one fourth or one half an inch in depth, and yet spread out to the width of a mile or more. (1894, 331)[16]

Because of the plethora of unclear and conflicting data, it is unwise at this point to present a definitive route (or routes) taken by the various Death Valley '49ers from Mount Misery to what we have determined to be Sheldon Young's "Misery Lake" in southern Coal Valley, Nevada (about a hundred trail miles northwest from Mount Misery).

Briefly, the problem is as follows: First, Manly provides contradictory information about where and when the Bennett Train turned west and where and

FIGURE 5.4. Bennett Springs on the mountain slope to Bennett Pass in the Chief Range west of Panaca, Nevada. This is the cold spring; there is a warm one also.

when the single men caught up to them. He says they all started north, but Brier and Young say it was west. He also does not differentiate between the Jayhawkers and the Bug Smashers, who were not always together. Second, journals of the 1858 Mormon exploring expedition that enlisted Asahel Bennett as their guide can be interpreted different ways as to where they picked up Bennett's '49er trail, and nothing is written about where they departed from it. It is not clear if the expedition trail applied to all the Death Valley '49ers or just to the Bennett Train. Third, Sheldon Young's diary gives some mileages without cardinal directions, and some of his travel has no mileages attached. Plus, his mileages, directions, and descriptions can be interpreted more than one way. And if Young's mileages are squeezed into the Bennett route described by Dame and Martineau, then some of Young's descriptions of the terrain do not fit their mileages. Thus, even with most (if not all) the historical data at hand and having been on the ground numerous times, attempts to create a single route that fits the available data and successfully passes scrutiny is like trying to force Cinderella's glass slipper onto her stepsister's foot.

Archaeologist Ed Caesar said, "The clues that remain will always prove insufficient to our curiosity.... What we know is always dwarfed by what we can never know" (2014). Richard Marius, an expert on researching history, reminds us that "an honest essay takes contrary evidence into account.... [When] different parts of the evidence contradict each other; using your own judgment about it all means that you must face such contradictions squarely" (1999, 24, 48). Thus, facing the contradictions squarely, I hereby refrain from the intense desire to provide a definitive detailed route from Mount Misery to Coal Valley in southwestern Nevada. Following, however, is an overview of this part of the route. Appendix E supplies data for those who wish to try to analyze it—a synopsis of the 1858 Mormon exploring route following '49er Asahel Bennett's guidance, Sheldon Young's diary for this part of the route, a list of Manly's varied inconsistencies, and descriptions and comments by other '49ers.

Overview: Route from Mount Misery to "Misery Lake" in Southern Coal Valley, Southern Nevada

All the Death Valley '49ers crossed the Cedar Range at some point and entered Meadow Valley, south of today's Panaca, Nevada. They ascended the Chief Range to Bennett Springs and crossed Bennett Pass (both named for '49er Asahel Bennett by the 1858 exploring company).

The '49ers then descended into and traveled north in Dry Lake Valley and then crossed the North Pahroc Range into White River canyon (or wash). Here they turned either north or south in White River wash to circumvent the Seaman Range before them. If they turned south immediately upon entering White River wash, they traveled down the wash and entered Coal Valley near its southern end. If they traveled north in White River wash, they passed the north end of the Seaman Range and then turned south and traversed Coal Valley from north to south.[17]

At some point before turning south, disagreements broke out among the '49ers. As John Groscup explains in his colorful way:

> Heare is whare the trubel began, and I never shall fore get that at the head of the valley whare wee ware campet before wee turned south in the grate American Deasert, and I never shall foreget the plase whare the counsial was held, which way wee should go, to the North ore to the south. and then A reconnoiter party was sent out which is as follows: Thomas Shannon, Wilyam Reude, Elic [Alex] Palmer, & my self. wee took the large spy glas that the Companey had.... [W]ee went up A very hy Mountain and with the aid of the spy glas wee could see A grate distence and alchalye [alkali]. (1885)

This range could have been the Seaman Range, or, if they had turned north as far as White River Valley, it could have been part of the Gwinn Canyon Range.

James Brier also alludes to the above disagreements when the company turned southward, "contrary to my advice" (1886). Some may have thought the high, snowy peaks ahead were the Sierra Nevada and their route to Walker Pass should curve to the south of them.

Misery Lake, Coal Valley, Southern Nevada

Locating Sheldon Young's Misery Lake was figured by using his diary to backtrack from the Franklin Well area on the Amargosa River where water and wood were available (a verifiable location) and by considering diverse routes from Mount Misery.

James Brier commented, "Our first troubles began...when we arrived at a big apparent lake, only to find it glazed mud, with a little alkali water. Hitherto we had had feed and water in plenty, but thenceforth they were rare" (1886). His highly compressed and garbled account is no help in finding the route, since his description could apply to other playas. As Manly said, there were several playas that tempted the '49ers.

John Brier, James's son, appears to refer to the Coal Valley playa when he wrote: "Descended into the first real desert I had ever seen, and saw here, for the first time, the mirage. We had been without water for twenty-four hours, when suddenly there broke into view to the south a splendid sheet of water, which all of us believed was Owen's Lake. As we hurried towards it the vision faded, and near midnight we halted on the rim of a basin of mud, with a shallow pool of brine" (1903, 331).

FIGURE 5.5. Playa, southern Coal Valley looking southwest to Murphy Gap separating Irish and Golden Gate ranges, 2016. Ranchers bulldozed the berm at left for Murphy Gap Reservoir. Fence controls cattle.

They were about 170 air miles northeast of Owen's Lake, with several valleys and mountain ranges between. The '49ers had no concept about the immensity of the western landscape nor how complex its topography.

Juliet Brier also referred to a miserable lake: "Instead of water we found it to be merely glazed mud covered with shallow alkaline water. From here on food and water were scarce."[18]

Charles Mecum seemed to speak of the same lake: "I cannot place my mind upon anything of note till we reached that warn [warm] stream and plenty of grass. stopt there 2 or 3 days to recruit. from there to wher we saw as we supposed quite a large Lake and drove to it in the eavening. used water that night, got up the next morning and all was dry about us as a powderhouse for miles around" (1872).[19]

This first mirage, or "Misery Lake," was on the playa in southern Coal Valley, probably at today's Murphy Gap reservoir, where ranchers have bulldozed an earthen berm to create a catch basin for rain and runoff from higher ground. Visible southwest from the reservoir is a tempting shallow notch called Murphy Gap at the south end of the Golden Gate Range that borders the west of Coal Valley.

Sheldon Young's "12m S.W" from the reservoir puts them about three miles beyond the gap in Cold Springs Wash. Here they lingered, confused about where to go next. Based on Young's "S course," their trail was in Wild Horse Valley, along Cold Springs Wash. They had an Indian guide who led them over the hills (a northern extension of Timpahute Range) into Sand Spring Valley. Starting at Misery Lake in southern Coal Valley, Young says:

[Nov.] 24) this Day lay in camp • the most of the Day exploring

[Nov.] 25) this [day] went 12m [miles] S.W course [to Cold Spring Wash] • had a Dry camp • it looks rather Dubious for water • misery lake is highly charged with alkali

[Nov.] 26) this Day lay in camp • caught two piutes • found that we were to[o] far north • let one of them go again & kept the other for a guide • caught another in evening

[Nov.] 27) lay in camp waiting for some of our men to come into camp

[Nov.] 28) this Day went 16m • had good roads • had a Dry camp & no grass • made a S course [Wild Horse Valley south of Golden Gate Range and then west through the Timpahute Range]

[Nov.] 29) this Day went 6m W to a spring [Sand Spring in Sand Spring Valley?] • had wood & grass • a Damnd Dubious looking country

[Nov.] 30) & the [Young was interrupted midsentence. He may have remained in camp or traveled this day, but we attribute him with zero miles traveled.]

Dec. 1) in camp Shoeing oxen • left in the evening • went 6m & camp • had a Dry camp • grass

What was the question the Jayhawkers asked for the response to be "we were to[o] far north"? Was it "how far" to water, to Ward's trail, to horses, to Owens Lake, to Walker Pass? "The clues that remain will always prove insufficient to our curiosity."

By the time the '49ers reached the southwestern part of Nevada, they were forced to ration what remained of their foodstuffs, and their cattle were becoming weaker each day. "Mrs. Bennett and Mrs. Arcane were in heart-rending distress. The four children were crying for water but there was not a drop to give them, and none could be reached before some time next day. The mothers were nearly crazy, for they expected the children would choke with thirst and die in their arms" (Manly 1894, 129).

Manly continues with another foreboding episode in this desert area:

When Bennett retired that night he put on a camp kettle of the fresh beef.... After an hour or so Mr. Bennett went out to replenish the fire and see how the cooking was coming on, and...he found that to his disappointment, most of the meat was gone.... We thought we knew the right man, but were not sure.... It is a sort of unwritten law that in parties such as ours, he who steals provisions forfeits his life. We knew we must keep watch.

John Colton gives a sense of their situation as they traveled through southern Nevada:

We thought we could make a short cut by crossing the Great American desert.... It was but a few days before we began to realize that we were in a bad fix. Water suddenly became very scarce, and such as there was was strongly impregnated with alkali or nitre. Only at long intervals did we find a drop of fresh water. Vegetation, too, had almost disappeared. There was nothing left but bitter sage brush and upon that our cattle had to subsist. One after another the beasts, starved and worn out, dropped down on the sandy plain and died. Our wagons were shortened to carts and we left the discarded wheels scattered along the plain with the carcasses of the cattle to mark for many years our terrible journey. (1888)

West of today's Rachel, Nevada, in southwest Sand Spring Valley, where the Belted Range forced the '49ers to turn south, Young's diary continues:

large vally [Emigrant Valley] • found plenty of puddle water standing grass [&] greese wood here • we found a shrub called the Spanish Bayonet • S course

[Dec.] 3) this Day had harsh gr[a]vely roads not much grass p[l]enty of water went 16m had a Dry camp S by W course

Somewhere along the trail the emigrants could see the highest snowy peaks in the distant Panamint Range with a low place north of them they named "Martin's Pass," for Jim Martin, one of the Bug Smashers. It became their next goal.[20] Later, Manly commented on the pass while he was at Indian Springs (forty-five miles northwest of Las Vegas):

In a due west course from me was the high peak we had been looking at for a month [Telescope Peak], and lowest place was on the north side [today's Towne Pass], which we had named Martin's Pass and *had been trying so long to reach*.... West and across the desert waste, extending to the foot of a low black range of mountains [Amargosa Range], through which there seemed to be no pass, the distant snowy peak lay still farther on, with Martin's pass over it still a long way off though *we had been steering toward it for a month*. (Manly 1894, 123, 135; emphasis added)[21]

The Jayhawkers are now near Groom (dry) Lake in today's government-restricted Area 51, but since Young does not mention going to a lake, I suspect their previous bad experience plus finding "plenty of puddle water" encouraged some of them to stay well west of Groom Lake. However, this does not preclude other '49ers from having gone to Groom or Papoose playas looking for potable water.

In this vicinity, the loose train of emigrants took three separate routes. The

Jayhawkers headed for the hills (to the west), hoping to find grass and water. The Brier family and Bug Smashers, led by James Martin and Captain Townes, formed another group that found Cane Spring (southwest), where they remained several days. The Bennett Train (seven to nine wagons) followed Lewis Manly's advice and continued south as far as today's Indian Wells, Nevada. (The assumption is the Wade family followed the tracks of the Bennett Train.) However, each of these discrete groups ended up at Travertine Springs on the eastern edge of Death Valley, although not at the same time. They were still aiming for Walker Pass, but they were now controlled by the dictates of the terrain—mountains, rough canyons, feed, and water.

Open to various interpretations is the exact route of the Jayhawkers through the desert valleys and canyons south of the Belted Range and north of Skull Mountain (in the atomic bomb testing area and today's Nellis Air Force Range). Based on Sheldon Young's diary of December 5–9, I choose the following route.[22] There is no way of knowing if all the Jayhawkers took this route; however, the Briers, Bug Smashers, and the Bennett Train did not.

Sheldon Young's (and Probably the Jayhawkers') Route from Emigrant Valley into Death Valley

Sheldon probably crossed Groom Pass, then turned south in Yucca Flat for a few miles. On December 6 he traveled only five miles, possibly because they found grass in Yucca Flat where some snow fell. From other accounts, there was not enough snow to provide water for the cattle. On December 7 they turned west toward Tippipah Point, circled around Mine Mountain into Mid Valley, and followed it south to Barren Spot, with Lookout Peak dead ahead. Here they veered southwest across Jackass Pass into Jackass Flats, possibly camping at the northwest tip of Little Skull Mountain. They continued west four miles to Fortymile Wash, where again it snowed and they found "a little grass." On December 10 they had "plenty of snow." This is the snowfall the Jayhawkers credit with saving their lives. Colton wrote Palmer March 21, 1894, "[We] were 5 days at a time, several times, without water."

From Emigrant Valley in the vicinity of Groom Lake, Young records:

[Dec.] 4) had one ox Driven off by the Piutes last night · had bad roads gravelly roads SW went 16 m
[Dec.] 5) rough road S course · found water not much grass had a Dry camp went 14 m [possibly over Groom Pass into Yucca Flat]
[Dec.] 6) went 5m Snowed · campt on Dry Diggins [Yucca Flat] · grass plenty no water · S course

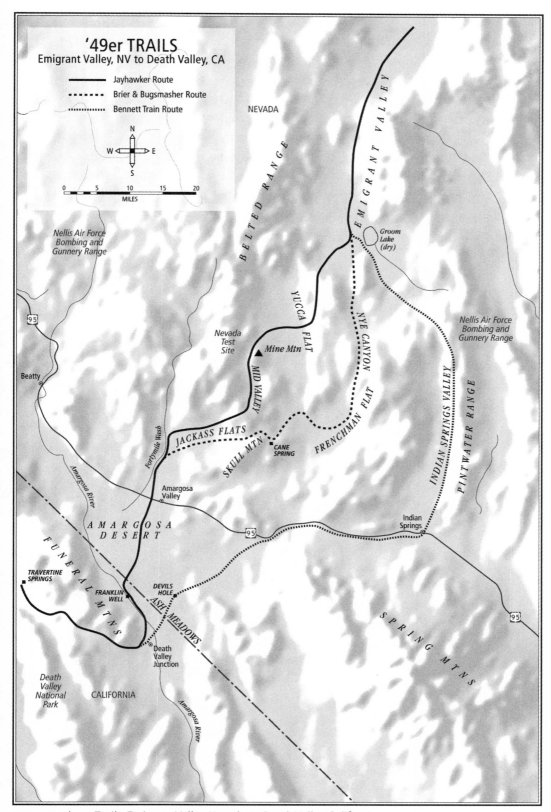

FIGURE 5.6. '49er Trails: Emigrant Valley, Nevada, to Death Valley, California

[Dec.] 7) went 12m had hard roads gravelly roads · grass verry scarce no
water · Damed hard looking Country · plenty of large mountains in
sight ahead· W course[23]

[Dec.] 8) this Day Bore S W · had bad roads no grass nor water · had a dry
camp · goin 12 m [into Jackass Flats]

[Dec.] 9) it snowd · found a little grass no water · went 4m W

[Dec.] 10) hard roads · plenty of snow [Fortymile Wash] · no grass nor
water · WSW course went 10m · had a Dry camp [in the Amargosa
Desert]

Years later, after talking with some of the Jayhawkers, Manly related their
lifesaving experience in this area:

As they stood around the fire a stray cloud appeared and hid the stars,
and shortly after began to unload a cargo of snow it carried. They spread
out every blanket, and brushed up every bit they could from the smooth
places, kindled a little fire of brush under the camp kettles and melted
all the snow all of them could gather, besides filling their mouths as fast
as ever they could, hoping that it would fall in sufficient quantities to
satisfy themselves and the oxen, and quench their dreadful thirst. Slowly
the cloud moved scattering the snowflakes till they felt relieved. The last
time the Author conversed with a member of this party was in 1892, and

FIGURE 5.7. From mouth of Fortymile Wash looking northeast. Jayhawkers, Briers,
and Bug Smashers came down this part of the wash.

it was conceded that this storm saved the lives of both man and beast in that little band of Jayhawkers. (1894, 333)[24]

Fortymile Wash empties into the Amargosa Desert, and the Jayhawkers crossed the desert following the water drainages southward to the foot of the Funeral Mountains. According to Sheldon Young:

[Dec.] 11) crossed a Desert [Amargosa Desert] 12m found wood grass & water [Amargosa River] Just in time to save our cattle · a number of our cattle gave out before getting them to water SW

[Dec.] 12) went 3m Down stream [near today's Franklin Well] · found better grass · lay in camp.

A short stretch of the Amargosa River (west of today's Longstreet Inn and Casino at the Nevada-California border) provides the most water along the river in this part of the desert, and there is plenty of wood available to create a forge to shoe the oxen. This area had been a major prehistoric Indian habitation site. Anthropologist Alice Hunt found a concentration of Indian sites scattered about a mile northwest of Franklin Well along the Amargosa River, sites dating as far back as two thousand years (Hunt and Hunt 1964).

Here the Jayhawkers rested, cut some of their wagons into carts, and sent scouts out to determine how best to cross the Funeral Mountains to Martin's Pass in the big snowy range still to the west. They stayed for six days while men and animals recuperated in what Sheldon Young called Relief Valley. Young:

[Dec.] 13th lay in camp shoeing oxen [near today's Franklin Well] · getting short of provisions · killed two oxen cool weather

[Dec.] 14 & 15 & 16–17) continued shoeing our oxen & exploring · found a hard looking country around us

[Dec.] 18) lay in camp shoeing oxen · killed two oxen [belonging to] Carter & Meekum [Charles Mecum] · Explorers all returned

[Dec.] 19) this Day rolled out of camp had soft roads in this vally · [it] is highly charged with alkali some blue Ash [and] greesewood[25] · went 8m S [to the southeastern tip of Funeral Mountains]

[Dec.] 20) this Day left relief vally · Bore a W course over a Barren Desert · Discovered an old pack trail supposed to be [Barney] wards · went 12m · campt without water or feed · W [toward Furnace Creek Wash]

[Dec.] 21) this Day started verry e[a]rly · had bad roads & a Dismal looking Desert a head [Death Valley] · no sighns of feed or water · there is a

FIGURE 5.8. Franklin Well area, Amargosa Desert, eastern foot of the Funeral Range where Jayhawkers lingered several days to rest and shoe oxen.

heavy range of snow mountains ahead [Panamint Range] · went 2m grass & water W [camp west of Navel Spring][26]

[Dec.] 22) went 6m · found plenty of grass & springs of water [Travertine Springs] · some horse signs

[Dec.] 23) lay in camp · killed ox · spent the Day in shoeing

The "old pack trail" Young saw was probably made by local Indians driving stolen horses from today's Tecopa area to their village in Furnace Creek Wash to use for food, not to ride. Young, and presumably others, was convinced they were still on Barney Ward's cutoff to Walker Pass.

The men circled Pyramid Peak at the south end of the Funeral Mountains, ascended the *bajada*, a sloping plain to the pass between the Amargosa Desert and Death Valley (route of today's Highway 190), and camped near the narrows where the Funeral Mountains and the Greenwater Range meet.

West from camp at the narrows, they found the Indian trail (still clear in the 1990s) to good water both at Navel Spring and just west of it, where Manly found "the Indians had made a clay bowl and fastened it to the wall so that it would collect and retain about a quart of water" at the foot of a "varigated clay formation" where "a little water seeped down its face" (1894, 137).[27] The wagons would have traveled in the wash downhill from the Indian trail. Here the Jayhawkers camped to water and graze the oxen while scouts went ahead. From this area Travertine Springs is about seven miles northwest.

The Brier Family and the Bug Smashers' Route

About December 5, in Emigrant Valley on today's Nellis Air Force Base (in Area 51), the three main parties (Jayhawkers, Bug Smashers and Briers, and the Bennett Train) each took different routes.

Due south of this separation camp (where the Jayhawkers turned west), Nye Canyon divides the Buried Hills from the Halfpint Range. In Nye Canyon the date *1849* is inscribed at the mouth of a side gully where there is a small *tinaya* called Triple Tanks (Long 1950, 224–25; Koenig 1984, 86–89). Margaret Long says there was a sign—"dig and find water"—nearby in the mid-1900s. Assuming the *1849* is authentic, it indicates '49ers (the Bug Smashers and Brier family) pioneered a wagons road through Nye Canyon to Frenchman Flat, where they connected to Cane Spring Wash and followed it westward to Cane Spring.

Cane Spring, nestled at the northeastern base of Skull Mountain, became a recruiting stop for the Bug Smashers and Brier family. James Brier described it as "a branch of Forty-Mile Cañon, where a foot of snow fell on us. We staid here a week to recuperate our cattle, suffering much from cold. Finding that the oxen would carry packs well, the company packed all necessaries on the cattle and burned everything else, with the wagons" (1886).[28]

The Southern Paiute and Western Shoshone intermingled in the Amargosa area as far east as Cane Spring. The residential base at the spring was called Hugwap, but the Southern Paiute called the winter camp there Pagambuhan. In fall, a large festival and rabbit drive in nearby Yucca Flat was attended by all surrounding families (Pippin 1986, 31–32). Willow and mesquite were prevalent at the spring, as were cane, cattail, Indian ricegrass, and several cacti and yucca. There would have been small fields of corn and melons, Indian homes, burial sites, shards of pottery and chips from arrowheads, scrapers, knives, and spear points (Stoffle et al. 1990, 127–33). The Mormons, who created a freight stop here, may have mined in the area as early as 1853, and the springwater was later piped to the 1927 mining camp of Wahmonie (the Paiute word for "yellow gold") (Quade and Tingley 2006, 7, 11). Much of an old stone stage-stop building still stands below Cane Spring.

Lewis Manly described the Brier–Bug Smasher camp as seen by a prospector about 1890:

> I saw a man Barton who 3 years ago went hunting the gunsight lead & he says he found an old camp a 100 miles east of Death valley. the stones whare you set your camp kettle still remain as you left them. here he found a silver dollar a[s] black as ink, a hard wo[o]d picket pin & some small wagon irons. he found a plac[e] whare the wagon & ox tracks ware

plain to be seen, as well as manure on the little brush. he says the squs [squaws] rooled the wagon tyre 75 miles to Pioche & got a big price for them. this camp is whare jim Martin left his wagons if your remember (you dont find this particular camp in the book). he [Barton] did not find the gunsight Lead that Lew West went back to locate.[29] (November 25, 1894)

John Brier recalled the abandonment of the family wagons: "the drifting sand, the cold blast from the north, the wind-beaten hill, the white tent, my lesson in the Testament, the burning of wagons as fuel, the forsaking of nearly every treasured thing, the packing of oxen, the melancholy departure" (1903, 333).

From E. B. MacBrair-Koller, a member of the Brier clan, we learn more about what the Briers took with them:

I do know that Juliet prepared a waterproof drawstring bag for the silver, and Columbus, according to my records, wrote in a letter, "We buried our treasures with great care thinking to return for them later."...I found a list of books the family had taken with them in the Conestoga wagon. "McGuffey's Reader, Pickes (?) Grammar and Murray's Geography, Swiss Family Robinson, Robinson Crusoe, Gulliver's Travels, Aesop's Fables, Adventures of Ulysses, Lamb's Tales of Shakespeare, and Richard Dana's Two Years Before the Mast. James Brier packed a large book of Commentaries, and his well worn Bible. I [Juliet] put in my favorite book of Poetry." (1988)

Juliet's silver and probably many of their books were buried near Cane Spring, not far from where atomic bombs were tested in the 1950s and what is now the Nevada Test Site. There the Briers and Bug Smashers abandoned their wagons and packed their meager belongings on the backs of their emaciated oxen.

In lower Fortymile Canyon they found the Jayhawker tracks and followed them southward across the Amargosa Desert to the Amargosa River. Whether they caught up with the Jayhawkers before the latter moved on is not clear, but they followed the Jayhawker trail over the pass into Furnace Creek Wash the day before Christmas.

Juliet gives a heartrending account of her trek into Death Valley:

We reached the top of the divide between Death and Ash valleys and, oh, what a desolate country we looked down into. The next morning we started down. The men said they could see what looked like springs out in the valley. Mr. Brier was always ahead to explore and find water, so I was left with our three boys to help bring up the cattle. We expected to reach the springs in a few hours and the men pushed ahead. I was sick and weary, and the hope of a good camping place was all that kept me up. Poor little Kirk[e] gave out and I carried him on my back, barely seeing where I was going, until he would say, "Mother, I can walk now." Poor little fellow! He would stumble on a little way over the salty marsh and sink down, crying, "I can't go any farther." Then I would carry him again, and soothe him as best I could.

Many times I felt I should faint, and as my strength departed I would sink on my knees. The boys would ask for water, but there was not a drop. Thus we staggered on over the salty wastes, trying to keep the company in view and hoping at every step to come to the springs. Oh, such a day! If we had stopped I knew the men would come back at night for us, but I didn't want to be thought a drag or a hindrance.

Night came down and we lost all track of those ahead. I would get down on my knees and look in the starlight for the ox tracks and then we would stumble on. There was not a sound and I didn't know whether we would ever reach camp or not. About midnight we came around a big rock and there was my husband at a small fire.

FIGURE 5.9. LeRoy Johnson at Travertine Springs, lower Furnace Creek Wash, Death Valley.

"Is this camp?" I asked.

"No; it's six miles farther," he said.

I was ready to drop and Kirk[e] was almost unconscious, moaning for a drink. Mr. Brier took him on his back and hastened to camp to save his little life. It was 3 o'clock Christmas morning when we reached the springs. I only wanted to sleep, but my husband said I must eat and drink or I would never wake up. Oh, such a horrible day and night!

We found hot and cold springs there [Travertine Springs] and washed and scrubbed and rested. That was a Christmas none could ever forget. (1898)

The Briers remained at Travertine Springs in lower Furnace Creek Wash after the Bug Smashers (or some of them) followed the tracks of the Jayhawkers to the valley floor and then northward toward their goal—Martin's Pass—to cross the snowy mountains before them. At Travertine Springs, Lewis Manly, who was scouting ahead of the Bennett Train, found the Brier family.

The Bennett Train Route (Southern Contingent of Death Valley '49ers)

The Bennett Train route from the Groom Lake–Papoose Lake area was down Indian Springs Valley (west of the Pintwater Mountains) to today's Indian Springs community, forty-five miles northwest of Las Vegas. They spent several days re-couping at the vacated Indian settlement while Manly climbed the western peaks to reconnoiter. Then the Bennett Train turned west (along today's Highway 99), past Point of Rocks, then southwest across the southern Amargosa Desert into Ash Meadows, where they camped at Collins Spring, three-quarters of a mile south of Devils Hole. Louis Nusbaumer commented that "at the entrance to the valley to the right is a hole in the rocks [Devils Hole] which contains magnificent warm water and in which Hadapp and I enjoyed an extremely refreshing bath" (December 23, 1849). As of 2016, the Indian-prospector trail was still visible near Devils Hole. The temperature of ninety-two to ninety-three degrees is well above Nusbaumer's estimate of seventy-five to seventy-nine degrees.

When the Bennett Train camped on the slope west of today's Death Valley Junction on Christmas Day, they "found the trail of the Jayhawkers going west, and thus we knew they had got safely across the great plain [Amargosa Desert]" (Manly 1894, 135).

All three groups—Jayhawkers and followers, Briers and Bug Smashers, and the Bennett Train—trudged into Death Valley within a five-day period via Furnace Creek Wash on their westward march.

CHAPTER 6

DEATH VALLEY

A trackless waste of sand and salt...enough to appall the stoutest heart.
—*New York World*, September 26, 1894

Colorful badlands, snow-covered peaks, beautiful sand dunes, rugged canyons.
—Death Valley National Park website, 2017

DEATH VALLEY IS A PLACE of mystery and intrigue. It has an aura that fascinates—lost treasure, poisonous water, slithering snakes, unbearable heat, vast silence, and dreadful death. As Richard Lingenfelter said, "The name inexorably transforms the land in the minds of those who hear it; no one can wholly ignore it" (1986, 7). Death Valley captures the imagination. As a natural wonder, along with the Grand Canyon and Yosemite Valley, it has become a major destination for visitors from around the world, but the mystique of its name adds a titillating dimension to its attraction.

"Death Valley" is the name William Lewis Manly gave the deep depression situated between the Amargosa Range on the east and the Panamint Range on the west. He named it when he stood on the crest of the Panamints on February 13, 1850, with Asahel Bennett and John Arcan. He named it to commemorate the death of a fellow comrade, Captain Richard J. A. Culverwell, whose desiccated body he and John Rogers found in the valley five days before. These men were part of the southern contingent (those who turned south in Death Valley) who became lost in the heart of the vast deserts of the Southwest and have since been called the Death Valley '49ers.

Over time this deep, narrow valley assumed a mystique of dread and abomination because of its name. The *New York World* cried out in 1894:

It is a pit of horrors—the haunt of all that is grim and ghoulish. Such animal and reptile life as infests this pest-hole is of ghastly shape, rancorous nature, and diabolically ugly. It breeds only noxious and venomous things. Its dead do not decompose, but are baked, blistered, and

FIGURE 6.1. Telescope Peak. Early morning looking west across waves of
dry salt on the Death Valley salt pan to water at Eagle Borax works below
Telescope Peak.

embalmed by the scorching heat through countless ages. It is surely the
nearest to a little hell upon earth that the whole wicked world can produce.

This valley was home for the indigenous residents—a small number of Shoshone
families, ancestors of today's Timbisha Shoshone Tribe.[1] It provided a warm valley to
live in during winter and cool mountains in summer. There were groves of mesquite
for beans and piñon trees for nuts. There were bighorn sheep, deer, rabbits, rodents,

and lizards for their arrows and traps as well as springs of warm and cold water—some fresh, some alkali. Although they lived a hard and austere life, they decorated their seemingly hostile landscape with geoglyphs, petroglyphs, and pictographs (rock alignments, symbols pecked into rocks, and pigment spread on rocks).

To describe the events and movements of the Death Valley Jayhawkers (and their fellow travelers), one needs to understand the north-south trending valleys—the Amargosa Desert and Death Valley, plus the Panamint, Searles, and Indian Wells Valleys—and the mountain ranges that separate them.[2] Therefore, I limit Death Valley, "where that bitter stream the Amargosa dies" (Lingenfelter 1986, vii), to all the land that drains directly into the valley: from the east, the slopes of the Amargosa Range that includes the Grapevine, Funeral, and Black Mountains, plus the Greenwater Range, and the Ibex Hills; on the west, from the Panamint Range that includes the Cottonwood Mountains.

A far broader definition of Death Valley was used by some of the '49ers and by various newspapers in the 1800s when recalling deaths that occurred in the California desert but not in today's Death Valley proper. The *Galesburg (IL) Republican Register*, for example, used an exceptionally broad description in 1895: "Just forty-five years ago yesterday, . . . thirty-two living skeletons suddenly rounded a sharp rock in the mountains. . . . Behind was the terrible American Desert, the 'Death Valley.'"

Although several '49ers other than Manly have been suggested for the honor of naming Death Valley—Juliet Brier, Sarah Bennett, Asahel Bennett, and

FIGURE 6.2. From Zabriskie Point, Death Valley, to Manly Beacon. Salt pan (*left*), Tucki Mountain in background.

FIGURE 6.3. Hills south of Furnace Creek Wash looking into southern Death Valley.

Nevada's first governor, Henry G. Blasdel—the data tips toward Lewis Manly as the person who deserves credit for naming it.[3] In 1877 a newspaper reporter interviewed Manly and wrote, "As they gazed back on the dreary waste from which they had just escaped, Mr. Manly exclaimed: 'There Lies Death Valley,' and it has borne that name ever since" (*Pioneer*, March 21, 28).

In his 1888 serialized account, Manly relates how he went to the crest of the Panamint Range with Asahel Bennett and John Arcan as the families left Death Valley for the California coast. "We stood here and talked a few minutes, and Bennett took off his hat, and Arcane and I did the same and said, 'Goodbye Death Valley'" (L. and J. Johnson 1987, 119). Had it been Arcan or Bennett, I think Manly would have given him credit.

In 1894 Manly published a similar description in *Death Valley in '49*: "We took off our hats and...spoke the thought uppermost saying:—'Good-bye, Death Valley!'" (216).

About 1998 Genny Hall Smith, author of *Deepest Valley: A Guide to Owens Valley*, contacted us after she read a first edition of Manly's book that belonged to her grandmother Mary Hall. In the back was this inscription: "William Manly gave Death Valley its name, 'Death Valley.' Mr & Mrs Manly were our good neighbors for many years, and Hayes Hall lived with them for a time when attending High school in Santa Clara. M. E. Hall." This inscription does not prove Manly named the valley, but it is another piece of evidence supporting that choice.[4]

The week ending November 23, 1860, Lewis Manly was staying in the Bella Union Hotel in Los Angeles at the same time as Joel H. Brooks, guide for the

California Boundary reconnaissance party (*Los Angeles Star*, November 24, December 1, 1860; L. and J. Johnson 1996, 16). Brooks had been a member of William Denton's survey team in 1856 and Henry Washington's team in 1857 when parts of Death Valley were surveyed for the General Land Office, but they did not attach a name to the valley on their maps (*Annual Report* 1857). There is no concrete proof Brooks and Manly conferred, but it is reasonable to assume they reminisced about the valley and that Manly likely told Brooks about the name. Four months later, an article about the ill-fated boundary reconnaissance ran in the *Visalia (CA) Delta* on April 13, 1861. They had aborted their northwest trajectory due to inferior mules, lack of feed, and poor water and were forced to traverse a "valley of great length...[that was] not less than 400 feet below the level of the sea.... This remarkable plain has been very appropriately named Death Valley."

Shortly after talking to Brooks, Manly with Asahel Bennett, William M. Stockton, and Caesar Twitchell returned to the Panamint Range on a rescue mission, which led to the first known appearance in print of the morbid moniker in the *Los Angeles Star* on March 16, 1861. Twitchell was interviewed for the article, "Mining News—Interesting from the Desert" after the men "rescued" Charles Alvord, who had been abandoned in the area by angry miners earlier that season (Manly in Woodward 1949, 61–79). While the rescuers were at Redlands Spring in the Panamint Range, Alvord blundered into their camp, and the four men continued to prospect. During this time Twitchell learned from Manly and Bennett about the valley immediately east of them and mentioned the name to the reporter: "At a place called 'Death Valley,' so named in 1849." The valley was actually named on February 13, 1850 (L. and J. Johnson 1987, 119n92; Appendix B). To give credit where credit is due, Asahel Bennett also has a good claim on being the first to pronounce the valley's name.

Ever since that distant winter of 1849–50, the valley called Death has taken, and continues to take, the lives of those who invade it without respect for its treacherous weather and deceiving distances (Cronkhite 1981). Now enclosed in a national park, this unique valley belongs to the people of our country (with the exception of lands belonging to the Timbisha Shoshone Tribe).

The 1894 vision of Death Valley continued in the *New York World* with more imagination than accuracy, but it is decidedly dramatic:

> Of fresh water there is none, but a liquid having the appearance of water oozes in some parts from the salt and lava beds. It is deadly poison. There is no humidity; simply a straightforward, business-like, independent, sizzling heat which keeps the thermometer in the region of 130 in the shade all the time, and the visitor has to provide his own shade....

[It is] a trackless waste of sand and salt, shimmering by day beneath the rays of a more than tropical sun. On entering there is no escape. Hemmed in on all sides by titanic rocks and majestic mountains, full of treacherous pitfalls, false surfaces, and quivering quicksands, surrounded by a silence that can almost be felt, in its most cheerful aspect Death Valley is enough to appall the stoutest heart.... But when night comes countless lizards squirm out of their burrows, rattlesnakes wriggle across the alkali crust, horned toads creep about and scorpions and tarantulas of enormous size sharpen their claws and bustle around in search of prey. (J. Johnson 2008, 3–12)

The Death Valley National Park website gives a more accurate description:

Hottest, Driest, Lowest

In this below-sea-level basin, steady drought and record summer heat make Death Valley a land of extremes. Yet, each extreme has a striking contrast. Towering peaks are frosted with winter snow. Rare rainstorms bring vast fields of wildflowers. Lush oases harbor tiny fish and refuge for wildlife and humans. Despite its morbid name, a great diversity of life survives in Death Valley.... Here you can find colorful badlands, snow-covered peaks, beautiful sand dunes, rugged canyons, the driest and lowest spot in North America, and the hottest in the world.

We have camped on the valley floor when water froze in our canteens, and on a spring day we have been caught hiking near Salt Creek when the temperature climbed to 111 degrees Fahrenheit.[5] Death Valley is indeed a land of extremes.

FIGURE 6.4. Brier family and Jayhawkers circled from right to left behind these dunes, then climbed the Panamint Range to exit Death Valley.

CHAPTER 7

DEATH VALLEY TO SEARLES VALLEY

If you're going through hell, keep going.

—Attributed to Winston Churchill

THE DEATH VALLEY '49ERS included about eighty men, four women, and eleven children (listed in Appendix B).[1] All of them funneled down Furnace Creek Wash, although not at the same time, and camped at Travertine Springs in lower Furnace Creek Wash on the east side of Death Valley.

Travertine Springs was part of a major Indian encampment that included the site at today's Furnace Creek Inn where "Kitchen Rock" and remnants of a large cave are still visible. A giant boulder is imbedded in the wash (part of the inn parking lot) with a dozen deep seed-grinding holes in the top—*matates*—that attest to a long period of occupation. A portion of the rock has been accidently bulldozed away, and past attempts to protect it with a stone wall have been covered with alluvium from torrential gullywashers.

To supplement their local food sources, the Death Valley Indians most likely stole horses along the Old Spanish Trail about sixty miles southeast of Death Valley. This was much closer than the San Joaquin or Santa Clarita Valley, each 260 miles away. A pictograph in the Greenwater Range (east of Death Valley) shows a man with a tall hat riding a horse (or mule), implying the Indians saw drovers along the Old Spanish Trail. Frémont and Kit Carson document an Indian horse raid at Resting Spring in April 1844 (east of today's Tecopa, California) (Frémont 1970, 677–78; Carson 1966, 82–85). When Sheldon Young saw "some horse signs," it was probably bones or dung from horses driven to Indian villages in Death Valley (December 20, 22, 1849).

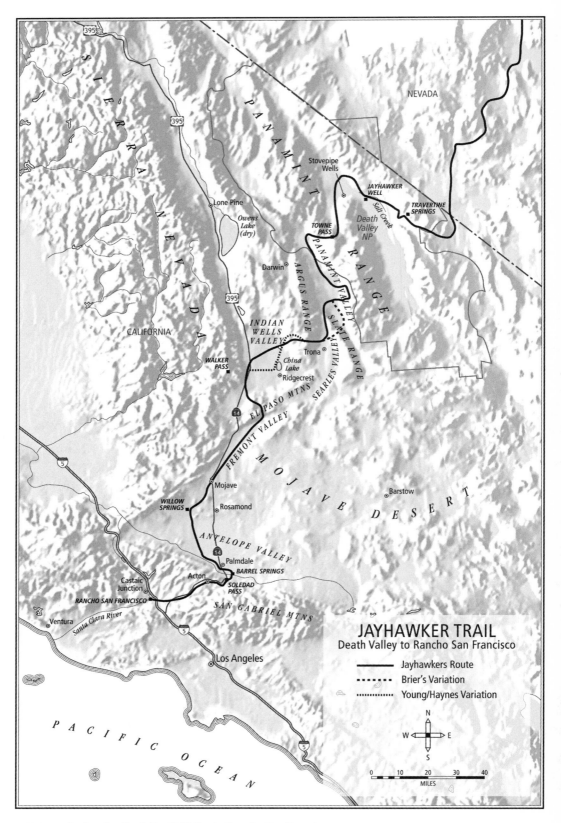

FIGURE 7.1. *Jayhawker Trail: Death Valley to Rancho San Francisco.*

Christmas in Death Valley, 1849

The day before Christmas, the Jayhawkers left Travertine Springs and followed Furnace Creek past clumps of mesquite, through the vacated Indian village, and into Death Valley proper. Colton recalled, "On Christmas day we struck a hollow full of boulders of rock salt and the hillsides were covered with the finest salt you ever saw" (1888). He wrote T. S. Palmer, "We cut our waggons down to carts about half way across [the Nevada desert] & Christmas morning we cut them into pack saddles, & our cattle toted the blankets & Jerk" (March 21, 1894).

The Jayhawkers confronted the rough, jagged salt pan west of today's Furnace Creek Ranch and, finding it impassable for their wagons and carts, returned to the east side of Death Valley, where they turned north. They followed the Indian trail when they could, then skirted the Mustard Hills (named for their color) to Cow Creek, which was watered by Nevares and other springs above the Park Service maintenance and residential areas. Parts of this Indian trail are still visible. Some of the men probably took their cattle up to the springs, looking for forage near the old park headquarters of today. Other Jayhawkers (including Colton) appear to have continued north on Christmas Day and may have camped at Poison Springs (today's Salt Springs) or continued to *Tugumu* (Indian name for "place in the sand"), a small, brackish hole about twelve miles north of Cow Creek now called Jayhawker Well.

FIGURE 7.2. "Kitchen Rock" in middle of parking lot at Furnace Creek Inn, mouth of Furnace Creek Wash. Indian women ground mesquite beans and grass seeds in the bedrock *metates* (holes).

FIGURE 7.3. Remnants of cave in fanglomerate cliff, possibly used as an Indian abode many years ago. Note inn wall on top of the cliff.

Sheldon Young outlined the Jayhawkers' travel in Death Valley after leaving Travertine Springs:

> [Dec.] 24) this Day went 3m out of a large Kenion into an alkali vally [Death Valley proper] · found we could not cross · returned to the same side again · made 4m NW to water & gras [Cow Creek area] · this has been a verry warm Day · there is honey locust here [mesquite]
>
> [Dec.] 25) Christmas Day · it has Been very warm · we are in a vally of salt & saleratys · some springs of fresh water putting in on the East side of the vally [east of Park Village and old headquarters area] · made 8m N.W · in night went 1m²
>
> [Dec.] 26) went 8m to the head of Salt vally · found a small stream of Salt water · no grass nor fresh water NW wen[t] 4m from there up this vally & campt found grass & water [*Tugumu*] (Levy 1969).[3]
>
> [Dec.] 27) came to the conclusion that our only alternative was to leave our waggons & carts & pack our oxen

In 1861 the U.S. and California Boundary Commission reconnaissance party found traces of *two* sites where the Jayhawkers abandoned their wagons, the first one near Cow Creek, the second at Jayhawker Well. Aaron Van Dorn, assistant surveyor, wrote:

> Riding [north from Furnace Creek] along the border of the plain, we observed the faint tracks of the emigrant wagons of '49 running along its margin, and this morning we came to a spot where they had camped [near Cow Creek], and which was evidently one of the scenes of their sad disaster. It was strewn with the relics of their wagons—the spokes, tires and hubs of the wheels, and the iron of the running gear, chain, broken pots and other remains of camp equipage marking the abandonment and destruction of part of the train.... [At Jayhawker Well he noted:] Here again we find the traces of the emigrant encampment, bones of cattle in plenty, and the less perishable parts of their abandoned wagons, trace chains, broken pots, etc. ([Van Dorn] March 11, 1861, in Woodward 1961, 52, 53 and in *Sacramento Daily Union* July 13, 1861).[4]

The '49ers made no specific mention of abandoning wagons in the Cow Creek area, but having two sites helps explain certain confusion in some of the post-'49er literature. For instance, John Spears has a picture looking north over Harmony Borax works with the caption "Death Valley: Near Emigrants' Last Camp" (1892, 18).[5] This is about fourteen miles southeast of the Jayhawker Well abandonment site north of Salt Creek and more than a mile from Cow Creek.

Nevares Springs, headwater for Cow Creek, is a complex of springs two miles east of Park Village named for Adolphe Joseph Nevares. He was a teamster-prospector from San Bernardino who claimed the springs in 1871. This would have been the best site for browse for the emigrant oxen; however, a few springs are closer to the valley floor, one being near the current park curatorial center—possibly Sheldon Young's "springs of fresh water putting in on the East side of the vally" (December 25, 1849). Cow Creek used to be a running stream, but the Nevares Springs are now captured for use by the U.S. Park Service administrative and residential areas. The creek bed is three miles north of today's Furnace Creek Ranch (Mendenhall 1909, 36; Palmer 2006, 20, 54).

While the Jayhawkers were searching for a way out of Death Valley, the Briers and Bug Smashers celebrated Christmas at Travertine Springs. Juliet Brier recalled:

> The men killed an ox and we had a Christmas dinner of fresh meat, black coffee and a very little bread. I had one small biscuit. You see, we were on short rations then and didn't know how long we would have to make provisions last. We didn't know we were in California. Nobody knew

what untold misery the morrow might bring, so there was no occasion for cheer.

Music and singing? My, no. We were too far gone for that. Nobody spoke very much, but I knew we were all thinking of home back East and all the cheer and good things there. Men would sit looking into the fire or stand gazing away silently over the mountains, and it was easy to read their thoughts. Poor fellows! Having no other woman there, I felt lonesome at times, but I was glad, too, that no other was there to suffer....

So ended I believe, the first Christmas ever celebrated in Death Valley. (1898; see also Belden 1954, 19–28)

Lewis Manly's "christmas eve meal was made of a big stake taken from a poor skinny ox you ware compelled to leave by your trail side (the only one not eaten entire that year) whare he died some days before" (February 10, 1893).[6] Manly was scouting for the Bennett Train in the vicinity of the pass from Ash Meadows into Death Valley (route of today's Highway 190). Late Christmas Day or the day after, Manly caught up to the Briers at Travertine Springs after the Bug Smashers (or some of them) had left. "I walked on down the lonesome dry bottom and about an hour after dark came to Mr. Brier's camp, where there were numerous weak springs and a little grass.... Mr. Brier started on in the morning" (Manly in L. and J. Johnson 1987, 56–57).

In his February 10, 1893, letter, Manly wrote about the same location but does not mention the Briers:

The next day [after?] christmas I came to one of your abandoned camps • here I found some bacon rinds you had thrown away [at Travertine Springs]. I gathered & eat evry one I could find & called them excellent.... [A]bout this time you ware in the north end of Death Valley preparing to leave your wagons & pack your oxen.

I now turned back to meet the teams [of the Bennett Train], which I found seven or eight miles up the canyon [Furnace Creek Wash].

On Christmas Day the Bennett Train was having a meager and doleful Christmas on the *bajada*, the wide, sloping incline above today's Death Valley Junction southwest of Ash Meadows. Among them was the Louis Nusbaumer mess—Hadapp, Culverwell, Fish, Isham, Gould, and Smith.[7] Nusbaumer wrote (in German):

The 25th December 1849. Christmas in the mountains of California. (Written at the campsite before leaving it.) This day was in general a day of sorrow for us, since the previously mentioned ox was no longer in a condition to go on and we were forced to throw away all our possessions in order to lighten the wagon enough that 1 yoke of oxen can pull it. I left here at the camp site 5 good linen shirts, boots, handkerchiefs, 1 basket, friend Adolph's hat, stockings, 1 buffalo hide, etc.... If we are able to reach the company which has gone ahead, we intend to load one ox with provisions and then to go further, if our strength, which is diminished through poor nourishment, permits us. But courage has not yet left us, and we hope with the help of God to still reach our destination.... I hope my dear wife has a happier Christmas day at home. (L. and J. Johnson 1987, 160–61)[8]

Incident at Indian Village (Mouth of Furnace Creek Wash)

When the Brier family left Travertine Springs to travel down Furnace Creek Wash after Christmas, they came to the evacuated Indian winter camp site in the wash at today's Furnace Creek Inn.

Manly recalled the camp had "the appearance of being continuously occupied by Indians" (1894, 139). Van Dorn said eleven years later that "a rancheria of over a hundred of them existed at the mouth of this creek" (in Woodward 1961, 49).

Here, both Manly and Juliet Brier commented on finding an old Indian, apparently left behind by the tribe. According to Manly:

When I got near the lower end of the cañon, there was a cliff on the north or right hand side which was perpendicular or perhaps a little overhanging, and at the base a cave which had the appearance of being continuously occupied by Indians. As I went on down I saw a very strange looking track upon the ground.... I followed the track till it led to the top of one of these small hills where a small well-like hole had been dug and in this excavation was a kind of Indian mummy curled up like a dog. He was not dead for I could see him move as he breathed, but his skin looked very much like the surface of a well dried venison ham. I should think by his looks he must be 200 or 300 years old.... He was evidently crippled.... I took a good long look at the wild creature and during all the time he never moved a muscle.... I now left him and went farther out into the lowest part of the valley. (1894, 139–40; see also in L. and J. Johnson 1987, 57–58)

Juliet Brier came upon the same old man:

As the company came up we found the thing to be an aged Indian lying on his back and buried in the sand—save his head. He was blind, shriveled and bald and looked like a mummy. He must have been one hundred and fifty years old. The men dug him out and gave him water and food. The poor fellow kept saying, "God bless pickaninnies!" Wherever had he learned that. His tribe must have fled ahead of us and as he couldn't travel he was left to die. (1898; see also Belden 1954, 19–28)

Juliet's account implies her family was traveling with some of the Bug Smashers. It is possible the Georgians and the Mississippians were no longer moving together and the Briers may have been more closely allied with Townes's mess.

After guiding the Bennett Train into Travertine Springs, Manly hiked north to find out what the Jayhawkers were doing: "I was at your camp the day after christmas whare you ware burning your wagons & was only in Dotys camp" (March 16, 1890). More likely, this was two days after Christmas when Manly arrived at Jayhawker Well "about nine o'clock at night," near today's McLean Spring, headwaters of Salt Creek. The Jayhawkers were abandoning all the remaining wagons and carts and making packsacks for both men and oxen. After spending the night, Manly "started back for our camp [at Travertine Springs] before daylight" with news that the Jayhawkers were burning their wagons to jerk the meat from their oxen, and the Briers and Bug Smashers were already on foot (L. and J. Johnson 1987, 58).

While Manly was checking on the Jayhawkers, back at Travertine Springs the Bennett Train was disintegrating. The Bennett and Arcan families (plus their four drovers and Manly and Rogers) were still a cohesive group, and the Wade family continued to camp alone. But Louis Nusbaumer's mess divided, leaving Nusbaumer, Culverwell, and Hadapp at Travertine Springs, while Fish, Isham, Gould, and Smith turned north. Manly mentions meeting some of these men on his return to Travertine Springs:

> Met Mr. Fish of Indiana and another man with their packs on their backs. They were both of our party and were nearly out of grub and their cattle being poor and weak they had given them to their traveling companion, Capt. Culverwell, and were going to try to make the balance of the journey on foot.... Mr. Fish had wound around him a beautiful long whiplash which he would not throw away. He said he might meet a chance to trade it for something to eat.... I thought of the poor prospects of these two old men ever seeing a settlement. (Ibid., 59; Manly 1894, 142)

The second man could have been Isham or Gould. Juliet said later in the trip that Gould "would pick up everything the rest threw away, until he had so much that Mr. Brier gave him an ox to carry his load" (Belden 1954, 26). Stephens said he was "the Old fellow" who "wore a plug hat" when he first joined them (1884).

Poison Spring

The only water on the route north between Cow Creek and the undrinkable water at Salt Creek is a small seep in the Furnace Creek fault about three miles north of Cow Creek wash. The seep is now labeled "Springs (Salt)" on topographic maps. In 1903 Gurley Jones responded to a letter from John Colton: "Yes that spring is called poison. it would make you sick in the condition you were in as the first time I went there I tryed it." Joel Brooks, guide for the boundary reconnaissance party in 1861, "succeeded in finding a spring [north of Cow Creek] and some grass; but drinking the water, it proved to be a powerful emetic, and vomited him severely. It was thought to be poisoned with antimony" ([Van Dorn], March 12, 1861, in Woodward 1961, 52).

Salt Creek, Jayhawker Well, *Tugumu*, and McLean Spring

About eight and a half miles northwest of Cow Creek is Salt Creek, home of a rare pupfish (*Cyprinodon salinus*) that grows to between one and two inches long, and Salt Creek is the site of the lowest-elevation "waterfalls" in the Western Hemisphere. In the creek is a pool used by ducks, coots, and other waterfowl where clumps of bird feathers attest to this being a hunting area for coyotes and kit foxes. Neither "Poison Spring" nor Salt Creek has potable (drinkable) water.

FIGURE 7.4. Salt Creek is not drinkable. The '49er wagons skirted the hills east of the creek to its headwaters, today's McLean Spring, where Jayhawkers dug a well nearby, hoping for better water.

When the northern contingent reached Salt Creek, wagons could not follow the Indian trail where it clings to the precipitous side of the bluffs. So the Jayhawks pioneered a wagon road around them to the east. This later became part of the original auto road whose bladed path you can see for miles.

The '49ers fortunately found barely potable water in a little well at a temporary Indian camp site called *Tugumu* (Levy 1969). The well is about two hundred feet east of today's McLean Spring, a primary water source for Salt Creek. In 1909 Walter Mendenhall said there were "two small wells" that were "brackish, but usable" (1909, 34 spring #24). These would be the little Indian well, undoubtedly enlarged and deepened by the '49ers (Jayhawker Well), and the boundary reconnaissance well dug nearby in 1861, but its water quality was no improvement (L. and J. Johnson 1996, 9; [Van Dorn] in Woodward 1961, 53). LeRoy has sampled water from Jayhawker Well and agrees with Mendenhall's assessment. On various old maps this site is labeled Jayhawker Well, Lost Wagon [site], or Salt Creek Wells. Jackass prospectors used them for another sixty years—their stomachs could tolerate almost anything. In the 1990s the well was about a yard wide and a yard deep with about a quart of water in the bottom and bunchgrass growing around it. Birds, coyotes, and kit foxes use it since it is the only source of "fresh" water for miles. A metal post stands nearby, but its sign has long since been stolen by souvenir hunters. It probably pointed the way west to the corrugated part of the road to Towne Pass and southeast to Furnace Creek. Or maybe it proclaimed mileage to the next water.

After the Jayhawks arrived at *Tugumu*, they realized their emaciated oxen could not pull wagons or carts up the steep *bajada* to cross the Panamint Range—the only way out was on foot. They slaughtered the weakest oxen and dismantled their wagons to get tinder-dry oak and hickory planks with which to make jerky from the ox meat. They were lost and stranded in the greatest void of uncharted land in today's contiguous United States. The only thing they knew for certain was the goldfields lay to the west. They remained here about three days (Haynes's diary says four) and undoubtedly sent scouts both west and northwest following Indian trails to look for water and a way out of the valley.

Aaron Van Dorn of the boundary reconnaissance party wrote the following description of the Jayhawker Well area:

> This end of the valley is enclosed by a chain of sand hills some two or three miles from the bottom, which the creek [Salt Creek] cuts through in a deep gap several hundred yards wide. All this end of the valley is very sandy, and the great body of the water evidently finds its way to the sink under ground, but in this gap a considerable portion of its waters

is collected in one channel and forms a series of wide, shallow pools of clear but intensely salt water. It is lined with salt grass and rushes, which harbor a few water fowl, principally teal ducks.

We continued up the creek some three miles from its mouth, keeping on the north [northeast] side in the sand hills to avoid the marsh of the bottom, and pitched camp in a mezquite grove on the flat near the creek, which is thinly covered with coarse salt grass, and contains a shallow well of very salt water [Jayhawker Well]. We dug a new well above the old one to try for purer water, but though the stream was very copious, the taste was equally saline. Its water was undoubtedly very injurious to the mules.

Here again we find the traces of the [Jayhawker-Brier] emigrant encampment, bones of cattle in plenty, and the less perishable parts of their abandoned wagons, trace chains, broken pots, etc. It was here, it is said, that they lost some of their companions by death, and nearly all their stock, and all their remaining wagons were left behind. (in Woodward 1961, 52–53)

No '49ers died at Jayhawker Well camp, but Van Dorn's newspaper article added fuel to the rumor that many emigrants died in Death Valley.

When Gurley Jones wrote John Colton in 1903 he provided another example of Death Valley hearsay mixed with personal experience:

The salt creek is just as you say. I got my drinking water about 4 miles from there by diging for it about 4 feet. there is one farly good spring in the salt creek but hard to find. Yes the sand has drifted a great deal and covered most of the white snowy place you mean [salt efflorescence]. I had to dig from 2 to 6 feet through the sand to find the old wagon irons. I am satisfied that is the place. there is an old, old Indian who tells me that the first white men who ever came in the country left there wagons there at this spot.

During the Jayhawks' stay at Jayhawker Well, the Bug Smashers, the Brier family, and several other men, including Fish, Isham, Gould, and Smith, arrived to enlarge the northern contingent. Although they were on foot, they would have followed the wagon tracks rather than the Indian trail through the bluffs. The trail, one person wide, separates from the wagon route near today's gravel road into Salt Creek. Along the bluff side, the trail narrows where soil has sloughed into it, and only the paws of coyotes and kit foxes keep it open now.

As Sheldon Young wrote, the Jayhawkers "came to the conclusion that our only alternative was to leave our waggons & carts & pack our oxen." The next day "the Bug Smashers came up with us packing their oxen • Still we are Busy making preparations for packing," and on December 29, "This Day Bug Smashers left."

During his brief stay at Jayhawker Well, Manly was told "the Martin party [Bug Smashers] was already in marching order and this camp was so poor that they did not wait, but gave all their oxen they had left to Mr. Brier" (1894, 337). James Brier said, "We caught up with the Jayhawkers, who were burning their wagons. There were about 75 of them—all single men. Here, too, the Mississippi [and Georgia] boys took their packs upon their backs, gave me their oxen (about 15) and went west" (1886).[9] Brier's "75" is a major exaggeration for the thirty-two or so Jayhawkers, but the number is not far off for all the northern contingent— Jayhawkers, Bug Smashers, Briers, Germans, Fish-Isham party, and teamsters who caught up later in the trek.

Routes from Jayhawker Well to Towne Pass

From Jayhawker Well, the '49ers took more than one route toward the low place in the mountains now called Towne Pass (their Martin's pass).

The Bug Smashers, the first to leave Jayhawker Well, doubtless followed the Indian trail west around the salt marsh that skirts the base of Tucki Mountain to climb the *bajada* toward the pass. The distance was about twenty-three air miles to their camp on Towne Pass and three-quarters of a mile straight up. (The elevation of Towne Pass summit is 4,962 feet, but their camp was about 4,000 feet elevation, and they started below sea level.) They continued toward the snow line on Towne Pass and camped below it for at least two nights. The mountain camp was about three miles east of the actual summit. This is based on data such as that of John Brier—"passed over a belt of snow" after leaving camp; Young's

FIGURE 7.5. Arrows point to Towne Pass (*middle distance*) and Juliet Spring (*right*). "Post" on right is the trunk of a dead palm tree. The '49ers' next goal is in sight— Towne pass over the Panamints.

mileage—"went 8m • Descended into another vally • campt at the mouth of the Kenyon"; and Colton—"The Georgia boys built signal fires to guide us to camp, and had melted plenty of snow for us to drink." A view of such signal fires, if located farther up the canyon, would be blocked by a thumb of the mountain for anyone approaching from the valley. From the 4,000-foot elevation, the pass trends southwest about five miles until it narrows to a descending canyon on the west side of the Panamint Range.

Eleven years later, another group of weary wanderers, the boundary reconnaissance party, ascended the Panamints from Jayhawker Well, following approximately the same route as the Bug Smashers. Aaron Van Dorn described their trek toward Towne Pass in 1861:

> There is a partial break in the Panamint mountains to the southwest of this camp, known as Town's Pass, and for this our route today was directed. Our course described a great semicircle, curving round a lofty spur of the mountains projecting into the valley from the south [Tucki Mountain], and entering a lateral valley which runs up southward to the Pass. Our march was along a constantly ascending grade.... Town's Pass...was directly before us as we turned south; but this [Towns pass] we left on our right, and entered the mountains to the southeast, up a large, wide arroyo coming from that direction [today's Emigrant Canyon], following up which for three miles, we camped at a very small hole of water...known as Hitchins' Spring. (Woodward 1961, 5; Lingenfelter 1986, 62; Doctor 1988; L. and J. Johnson 1992)

The reconnaissance was heading for Towne Pass, but it turned southeast into an unnamed canyon (today's Emigrant Canyon) for water at two springs known by two of its party, James Hitchens and J. H. Lillard, who had prospected there with Darwin French a few months before. Up the canyon from Hitchen's Spring is today's Upper Emigrant Spring that Dr. Owen called "Marble Spring" (1861).[10] As late as 1861 the canyon and its springs carried no connotation related to emigrant use.

Although Van Dorn does not refer to following an Indian trail, remnants of its looping course can be found along the northern base of Tucki Mountain. It is particularly clear near the mouth of Emigrant Canyon.

Back at Jayhawker Well on the afternoon of December 30, 1849, Sheldon Young reports the Jayhawkers "got our packs ready & started • left [at] 3 oclock went 12m • had Dry camp • Cattle acted like old hands at it." Next day they "went 24m • got to the [t]op of a high mountain • plenty of snow • Dry camp."

Neither Young nor Haynes gives a direction of travel from Jayhawker Well, but Young makes clear he and his Jayhawker companions did not camp at a spring or well after traveling twelve miles; they probably continued at night by the light of the full moon before camping. Had he followed the Indian trail that hugs Tucki Mountain as did the Bug Smashers, Young's total of twelve miles plus twenty-four miles from Jayhawker Well to the camp on Towne Pass is about twelve miles too far.

Manly suggests the Jayhawkers headed north before turning westward to Towne Pass: "The Doty party, or Jayhawkers, when they were ready, started first a northerly course.... They soon came to some good water [Juliet Spring?], and after refreshing themselves turned westward to cross the great mountain before them" (1894, 337).

Young's and Haynes's mileages (twelve and fifteen miles, respectively, then twenty-four to the summit camp) strongly suggest the Jayhawkers and Briers headed north following the Indian trail that links the sparse springs along the eastern edge of the sand dunes, a result of the Furnace Creek fault where underground water from the Grapevine Mountains is forced to the surface.[11] To the immediate west is Mesquite Flat, an expanse of sand sprinkled with low dunes and clumps of mesquite.

About seven miles north of Jayhawker Well, the view into Towne Pass (their Martin's Pass) became visible around the shoulder of Tucki Mountain. Nearby

FIGURE 7.6. Outflow from Juliet (old Palm Tree) Spring. Author, grandson Justin, son Eric, 2016.

is the pretty little spring LeRoy and I named Juliet Spring to commemorate the only woman in the northern contingent.[12] It is about a mile south of now dry Triangle Spring.

The route was now clear to the Jayhawkers: west across sandy Mesquite Flat and southwest up the *bajada* to the low point in the snowy mountains toward which they had been steering for several weeks—Martin's Pass—soon to be renamed "Town's Pass."

John Brier (who was six years old in 1850) described it this way: "Steering north of west we drove across the shifting dunes for twenty miles to the pass between the Panamints and Telescope Peak" (1903, 334). Today, the mountains on both sides of Towne Pass are included in the Panamint Range, but those on the north are also designated the Cottonwood Mountains. Telescope Peak is nineteen air miles southeast of Towne Pass. As the argonauts ascended the Panamints from Death Valley, they could see Pinto Peak (7,508 feet) guarding the pass on the south. Towne Peak (7,287 feet) is the highest point on the north but not visible from the valley.

Town, Town's (Townes), or Towne Pass

Crossing this mountain range loomed large in the minds of the Death Valley '49ers. Many thought the goldfields or even the Pacific Ocean were just over this mountain, and thus it had to be the great Sierra Nevada. Tom Shannon said, "When our party got to where we thought we should be able to see the pacific ocean from the tops of the highest peaks, Bill Rude, Alex Palmer and myself started from camp about midday determined to assend the peak, in full confidence we should see the pacific.... [T]wo boys reached the summit but saw no ocean" (1894).

Most of the northern contingent traversed the Panamint Range via Martin's Pass, well known by 1861 as Town or Town's Pass and today the major passage over the Panamint Range via Highway 190.

The pass was named for Paschal H. Townes, one of the Bug Smashers from Mississippi who started west with the Pontotoc and California Exploring Company; thus, it should be spelled "Townes Pass" (Hoshide 1996, 215; Thomason 1978, 23; Rasmussen 1994, 8). Little is known about Paschal Townes, but Jackson Thomason, who traveled with him from the Missouri River to Salt Lake City, recorded on May 16, 1849, "Towns is a man I think with two faces & courts the good will of all & will speak in hard terms of all behind their back but I hope I shall in the future find cause to change my opinion" (1978, 34).

According to James Brier, Martin's Pass was renamed "Towne's Pass" by the Brier-Townes group. "Thence we crossed the Telescope range by Towne's pass—named by us after our captain" (1886).[13]

Brier recalled his return trip of 1873: "The pass over is called Town's Pass. I had no difficulty in going directly to all these points although 23 years had elapsed.... At our old camp in Town's Pass where we melted snow I found a Butcher Knife. At most of these points I wept over the remembrance of our sufferings" (1876).

In 1883 Henry G. Hanks, a California state mineralogist, attributed the naming of Towne Pass to some prospectors:

> In May, 1860, a party of fifteen men, headed by Dr. Darwin French, left Butte County in search for the Gunsight lead, said to have been found by the emigrants. They journeyed via Visalia ... across the head of Panamint Valley, thence by a rocky pass to a camp in Death Valley, where the emigrants left their wagons [at Jayhawker Well].... The party returned by the way they came.... They became satisfied that a pass they came through was the same by which Towne and the saved of the emigrant party made their escape, which led them to name it "Towne's Pass." (33–34)

These prospectors included Hitchens and Lillard, who joined the boundary reconnaissance party a few months later (Lingenfelter 1986, 62).

Apparently, the pass was first mentioned in print in 1861 in the *Sacramento Daily Union* as "Town's Pass, a gap through the Panamint Mountains." That same year Minard Farley had "Town's Pass" labeled on his published map of newly discovered silver mines ([Van Dorn], March 13, 1861, in Woodward 1961 and in *Sacramento Daily Union*, July 31, 1861; Farley 1861).[14]

First to arrive at the eastern part of the pass were the Bug Smashers (or some of them).[15] John Colton recalled, "The Georgia boys built signal fires to guide us to camp, and had melted plenty of snow for us to drink" (November 20, 1895). The year before, he wrote Palmer, "The Georgians that packed on their backs from that point [Jayhawker Well] & gave us the bal[ance] of their cattle were old Silver miners in Georgia—and while they were preparing [at summit camp], we stayed with them 2 or 3 days. we had plenty of snow to make water" (April 12, 1894).[16]

I have lumped the Georgians, Mississippians, and others into "the Bug Smashers" for convenience because information about their finer divisions is scant. Three men within the Bug Smashers are mentioned as being a captain: Manly thought "Jim Martin was chosen captain," Brier called Towne "captain," and Edward Coker mentioned "Capt. Nat. Ward."[17] Brier referred mostly to the Mississippi boys, and Colton referred to the "Georgia boys." These subgroups within the Bug Smashers were becoming more distinct in Death Valley.

The *Kansas City Times* reported that "the Jayhawkers started out [to ascend the Panamint Range] expecting to reach them that night, but did not come up to the party—Georgians—until 3 o'clock on the second morning. They had been

FIGURE 7.7. General location of Towne Pass camp near the snow line. Pinto Peak is on the skyline near the 4,000-foot elevation sign.

all this time without water. The Georgians had snow ready melted for them" (February 5, 1890).

Young records two nights in camp on the pass:

[Dec.] 31) [From Mesquite Flat] went 24m got to the [t]op of a high mountain [Towne Pass] · plenty of snow · Dry camp

Jan 1 1850 this Day lay in camp · this Day had plenty of Snow to me[l]t for the cattle

[Jan.] 2) left camp at noon & went 8m · Descended into another vally [Panamint Valley] · campt at the mouth of the Kenyon · W course · lost one steer[18]

Young's "Dry camp" connotes he and his companions did not find a spring but camped not far from the snow line. His "top of a high mountain" refers to the pass over the range, not to the summit of a particular peak.

Juliet Brier recalled that from camp on Towne Pass, "the men climbed up to the snow and brought down all they could carry, frozen hard. Mr. Brier filled an old shirt and brought it to us. Some ate it white and hard and relished it as though it was flowing water, but enough was melted for our frenzied cattle and camp use" (1898). John Brier wrote, "The reunited company, now relieved of the cumbrance of wagons, steered their course for Town's Pass.... [There would be] no rest short of the snow-line. Snow lay in patches not far above the spot chosen for a camp; and quantities of it were brought down, in sheets, to be melted for the oxen" (1911a, 3).

The Bug Smashers had traveled for weeks with the Briers, and both Manly and James Brier say they gave their remaining oxen to the Brier family at Jayhawker Well. At some point Brier "lent" three or four oxen to the Jayhawkers and sold some to Doty (James Brier 1898; see also Manly, November 23, 1890). However, it is possible the Bug Smashers had taken their very strongest oxen with them and on the pass decided even those could not keep up, so a final jerking of ox meat was required—or these oxen were given away.

Towne Pass was used by a handful of prospectors from 1850 for at least ten years before the water-rich Wildrose Canyon route came into use. In 1925 Bob Eichbaum bulldozed a road across Towne Pass in the general vicinity of the northern contingent's route and started his resort at today's Stovepipe Wells, but that is another story.

We must consider that some men may have followed different routes over the Panamint Range. Some may have entered Emigrant or Jayhawker Canyon because Indian trails led to springs in those canyons. A trunk found near the top of Pinto Peak and an article about Jayhawker William Rude returning to Death Valley suggest additional routes could have been used.

The Pinto Peak Trunk

In 1999 Jerry Freeman found a little wooden trunk while he sought the Jayhawker route over the Panamints, a trunk someone placed in a small cave below the summit on the western flank of Pinto Peak. It was filled with a bizarre collection of vintage items and included a "Propity Manafest of/William Robinson Second/Day of the Lords Year 1850." The implication was this trunk belonged to Jayhawker William Robinson who died at or near Barrel Spring south of Palmdale almost a month after the Jayhawkers crossed the Panamints. The U.S. Park Service had a few items analyzed, and the results showed some were made after 1850, proving the trunk and its contents were not placed on Pinto Peak by a Jayhawker in early 1850. The hoax created a diverting furor for a short time, and the trunk now resides in the Curatorial Center in Death Valley. Whether it was a prank or planted to "prove" a particular analysis of the route will probably never be known (L. Johnson 1999, 252–77).[19]

Some of the Bug Smashers did climb to a high point to see the land ahead. John Brier, who garbles many location names in his accounts, said his family was "about to enter the Panamint Desert [Valley] and soon, at its northern extremity, they went into camp.... There the Town party reappeared, having climbed Telescope Mountain and made a detour involving much labor and loss of time" (1911a, 3). Of course, they did not climb Telescope Peak, many miles to the south, but they could have climbed prominent Pinto Peak south of their camp. Tom

Shannon says two Jayhawkers, Bill Rude and Alex Palmer, reached the summit of a peak (1894).

William Rude Returns to the Panamints

In 1869 William B. Rude (later spelling his name Rood, Roods, or Rhodes) returned to Death Valley with George Miller (and two other men) to search for the "Gun-sight mine" and the cache of gold coins the '49ers buried near the "Summit Camp" (Miller 1919, 56).

On this trek Rude had a pack animal to carry his gear, which would have included a prospector's pick, a single jack, and chisels. The Rude-Miller party prospected today's Emigrant, Jayhawker, and Lemoigne Canyons and the Towne Pass area—major contributors to the detritus in Emigrant Wash fan. Our contention is that Rude chiseled his two inscriptions on his 1869 trip, not on his exit route in 1849 (L. Johnson 2013, 285–91).[20] "W.B.R. 1849" is on a small boulder along the wash in Jayhawker Canyon, and "W.B. Roods 1849" is on a larger one along the Indian trail to Lemoigne Canyon northward from Emigrant and Jayhawker canyons.

Two pieces of data hint at the possibility that some of the Jayhawkers may have crossed the Panamints through Emigrant Canyon and over Emigrant Pass. One is the name "Emigrant Canyon." The second is based on George Miller's account written in 1919, fifty years after his single trip into the Panamints with William Rude.

Today's Emigrant Canyon and its springs had no name for more than ten years after the '49ers crossed the Panamints. There is no record of emigrant artifacts being found by prospectors, miners, or the boundary reconnaissance party in 1861. Dr. Owen called the upper spring Marble Spring, not Emigrant Spring. In 1875 Dr. S. G. George wrote to the *Panamint News* in 1875 about the new name for Marble Spring. George and his companions had prospected the Panamint Range in 1860 and 1861. At that time he passed the area where "Hunter and Co. are now working the Nellie Grant mine. We gave the spring there the name of Marble Spring. I am informed that it is now [in 1875] called Emigrant Spring." Neither the canyon nor the pass is labeled on Wheeler's 1877 map, but a spring in the vicinity, whose exact location is unclear, is named Emigrant Spring.

George Miller said, "Mr. Rhodes told me…[he camped in 1849] at the place now known as Summit Camp, or Emigrant Pass between Death Valley and the head of Panamint Valley." The only pass debouching into the "head of Panamint Valley" is Towne Pass, but the way Miller's report is written implies that Rude's party based their search for the coins in the Emigrant Canyon and Pass area. Miller also says, "This was the camp where the silver ore was brought in," and

"this camp at the Summit Pass was the last camp they all made together," implying the Bug Smashers and others were all there (1919, 56–57). If Miller's "Summit
Pass" is Emigrant Pass, much of the convincing data from several other '49ers
does not fit. Thus, I find Miller's account interesting but of questionable value.

A preponderance of data makes clear the Briers, Sheldon Young's mess,
Colton's group, Asa Haynes's mess, Townes's mess, and some of the Bug
Smashers crossed the Panamints via Towne Pass. The data includes Young's and
James Brier's mileages, Brier's description of "head of the [Panamint] valley,"
naming "Town's Pass" for one of the Bug Smashers, John Brier's description of
descending a canyon with the "sudden emergence" of Mt. Whitney and then
their camp at the "northern extremity" of Panamint Valley, and Bug Smashers
providing Colton and his companions melted snow, not springwater, after their
prolonged hike up the *bajada* from the valley floor. These examples describe
crossing the Panamints over Towne Pass.[21]

Five Thousand Dollars and Silver Mountain

The Georgians, John Colton noted,

> were practical miners. They were almost as much reduced as were we. Up
> to that point [crossing the Panamints] they had carried with them $5,000
> in gold. But seeing that it would be impossible to transport it further, they
> securely cached it on the mountains, thinking to return for it. The gold
> has never been found. In their wanderings across the desert they had, too,
> come upon a mountain where they ran across the richest silver mine any
> of them had ever seen. I remember that they had specimens with them
> which they pronounced fabulously rich. (*Jefferson Souvenir*, after February
> 4, 1896)

Manly also referred to the gold coins the Bug Smashers buried. In one letter he
said, "I cant quite think the Georgia boys carried $4000 so far, too much of a
burden," and in another, "The $5000 the Geo boys buried will never be found.
40 years burries evrything on that trail out of sight" (February 7, March 16, 1890).

George Miller's article said Rude told him that before "they separated at this
place [presumably the Bug Smashers going a higher route over the Panamints]
they divided up the money they had, each taking what he wanted, and dumped
the rest in a blanket, about $2,000.00 or $2,500.00, and buried it under a greasewood bush" (1919, 56).

One of the Bug Smashers picked up a piece of silver float as he climbed the
fan toward Towne Pass. The story has variations, but in general the ore was soft
enough to be fashioned into a new gunsight to replace a broken one. This chance

find started a cavalcade of prospectors hunting for the fabulous Lost Gunsight Load. Manly quipped, "Our trail in 49 has been looked over evry year, since we made the first white mans tracks on the Desrt, searching for the Gunsight Lead" (July 1894). "As to Silver Mountain," Colton wrote T. S. Palmer, "the Georgians... were old Silver miners in Georgia.... [T]hey frequently remarked of the great richness of the Silver ore around the camp & said it was fabulously rich—but would give it all to know how to get out—I dont think it has ever been found" (April 12, 1894).

Into Panamint Valley

John Wells Brier wrote a graphic description of their descent into Panamint Valley after leaving summit camp on Towne Pass, probably based on reports from his father who had returned to the area in 1873: "[We] passed over a belt of snow, of which the oxen greedily ate; then opened the winding passage way, walls, high and ever heightening on either side, a footing of sand, or jagged rock, often boulder-obstructed, precipitating leaps of difficulty and danger—a long trail of weary longing and rude discouragement of hope! A sudden emergence brought Mt. Whitney and the Minarets [the Needles] into glorious prominence" (1911a, 3).

Today's Highway 190 parallels the sandy streambed west from the summit of the pass until the mountain walls pinch it into an abrupt, boulder-filled canyon that drops well below the highway to a complacent path of sand.

FIGURE 7.8. Panamint Valley, Darwin Plateau, Sierra Nevada from Highway 190 looking northwest. Arrow points to the canyon the '49ers used to descend the Panamint Range. Rainbow Canyon is middle left; Lake Hill, middle right.

The Bug Smashers who climbed to a high point saw the Whitney massif of fourteen-thousand-foot peaks at least fifty miles ahead, and the California goldfields were on the other side of the mighty but still distant Sierra Nevada. It was a "rude discouragement" for those who descended the canyon and saw the "sudden emergence" of the Sierra Nevada, still far away.

Bugs Smashers Leave Panamint Valley

James Brier wrote for a Jayhawker reunion:

> You all remember the snow Mt that we crossed after you left your waggons. The name is Panamint, & the valley into which we descended on New Years day, is Panamint Valley. You recollect when some of the Mississippi boys, with little West, & Tom & Joe the darkies, left us & steered for a Cannon on the west side of the valley. The cannon is now called Darwin's, and at the mouth [upper end] of it, is a large mining town, called Darwin City. (1879)

Edward Coker (from New York) said twenty-one men escaped the desert over Walker Pass, but he names only eight, including himself: "Capt. Nat. Ward, Jim Woods, Jim Martin of Missouri, John D. Martin of Texas, 'Old Francis,' a French Canadian, Fred Carr, Negro 'Joe' and some others from Coffeeville, Miss, and others from other states" (in Manly 1994, 373; James Brier 1879; Evans 1945, 254–59).[22] William Osborn, who carved his name near Mount Misery, was also from the Pontotoc (Mississippi) and California Exploring Company. The name "Townsend" is often mentioned in the literature and may have been a man other than Paschal Townes.

In Panamint Valley the weary yet determined Bug Smashers plodded past the sand hills just east of today's Panamint Springs Resort as they followed an Indian trail into the mouth of Darwin Canyon, hoping to find lifesaving sweet water before the greedy sand swallowed it all. Ahead, in the bend of the wash, vegetation signaled moisture, although no water likely came to the surface, but three miles up the canyon was a true oasis in the desert. A short stream issued from a lovely fall, cascading into a shaded pool of water, today's Darwin Falls. After drinking their fill, they would have followed the Indian trail that swings away from the stream and up the canyon as it cuts through the Argus Range.

Somewhere west of today's former mining community of Darwin, the Bug Smashers came upon Indians who led them to water. Manly says it was Little Owen's Lake (today's Little Lake, twenty-five miles south of Owens Lake on Highway 395). The Indians may have led them to Walker Pass or through a shortcut to the Kern River as the sky turned gray and snow began to threaten with the approach of a severe Sierra storm (Horst 1996).

Manly described the Bug Smashers' trek west of Panamint Valley:

I became acquainted with Jim Woods, one of the [Jim Martin] party, at
Georgetown [north of Placerville] and worked by his side at the mines
near there, and from him I learned something of this party.

They were the first who left the wagons, and came through by little
Owen's Lake and across Walker's Pass, coming down Kern river on the
west side of the mountain. They came near starving; their meat was all
eaten up and they were trying to find game. The first they killed was a fox,
and, by saving every part, this lasted them till they got better game, and
they pushed on and reached the Mariposa mines. (March 1889)

George Evans, a prospector at Agua Fria (six miles west of Mariposa), recorded twelve men arriving at the Mariposa Mines on January 29, 1850, and ten
more February 9. The stories told by these men distinguish them as the Death
Valley Bug Smashers.

Jan. 31…On Tuesday last a party of twelve men by accident came in here,
in a starving condition.… The fragment now in hand reached Owen's
[Little] Lake and here met with a party of Indians who directed them to go
south along the mountain until they reached Walker's Pass and through
it pass to the west side. Following these directions, and subsisting upon
acorns, mule meat, and a few fish obtained from the Indians, they hap-
pily reached the mines without the loss of a man, after having suffered
intensely. These men think the other party about twenty days behind,
and what renders this more painful is the fact that there are women and
children suffering with those behind.

Feb. 10 (Sunday)…Ten men, belonging to the train mentioned as being
yet in the desert beyond the mountains arrived here last night, having left
the main body some three hundred miles from here, all in a starving, and
some in a perishing condition. The Indians met by these men were very
friendly and rendered every assistance in their power, and it is to them,
they owe the safety of their lives at this moment. (1945, 254, 259)[23]

James Brier confirmed: "They got through Darwin Cañon, into Tulare [Valley],
and at last to the Mariposa mines, after great suffering, but no loss of life" (1886).

The remaining Bug Smashers (part of the Mississippi boys)—Townes,
Crumpton, Mastin, and the Turner boys—traveled with the Briers down
Panamint Valley another thirty air miles to Post Office Spring, southeast of to-
day's ghost town of Ballarat (James Brier 1898).[24]

Down Panamint Valley

When the northern contingent emerged into Panamint Valley through the canyon just north of today's Highway 190, in front of them a wet playa stretched both north and south. After the Bug Smashers had headed west, James Brier described the moves of those remaining—the Jayhawkers, the Brier family, Townes's mess, and a few other men. "You all remember that we went down Panamint Valley 2 days journey when we reached the great Mesqueete Swamp and camped 2 or 3 days at a pool where there were a number of Indian Wickiups [Indian Ranch, about seven miles north of Ballarat]" (1896).

The argonauts arrived first at the Indian village. Some wickiups were in the dunes near (and probably at) today's Indian Ranch at the mouth of Hall Canyon closer to better water (a spring up the canyon). There were Indian trails to the rancheria, and leafless trees may have been visible, indicating water. Julian Steward said this was "the principal and probably only village within the northern part of the valley... called Ha:uta" (1938, 84, and fig. 7, village 42). He shows it as a winter village. Years later, Indian Ranch was owned by Panamint George Hanson. A mile and a half south is the "Mesqueete Swamp" that attracted ducks and geese—protein for the Indians.

There is a discrepancy between James Brier's and Sheldon Young's mileages from their camp in upper Panamint Valley to the Warm Sulphur Springs area. In 1876 Brier said, "From the Muskeet Swamp [Warm Sulphur Springs] to the head of the valley where the Mississippi Boys left us is 20 miles & 20 more to Death Valley." Three years later he said, "We traveled down Panamint Valley 25 miles to some Indian Springs." Young gives "24m S by E," then 4 miles east to the Indian encampment. (He also has an illegible mileage number during this part of the trek.) To account for the discrepancy between Young's longer mileages and Brier's 1873 mileages, I conclude the argonauts swung north around "plenty of water" and soft mud on the Panamint playa (the route the Bug Smashers would have followed to reach Darwin Canyon) before turning south. This detour was something Brier did not need to do upon his 1873 return trip as a guide for some miners.

Sheldon Young's diary in Panamint Valley also sheds light:

[Jan.] 2)...Descended into another vally [Panamint Valley] • campt at the mouth of the Kenyon • W course • lost one steer
[Jan.] 3) went on W to the Botom of the vally wholly Destitute of grass • plenty of water some greesewood [creosote bush] for Browse[25]
[Jan.] 4) went [*illegible*]m S Down the vally • water no grass Country looks hard ahead • cattle were fast a failing three was left today

[Jan.] 5) went 24m S by E • we have been 7 Days without seeing a bit of grass • had a Dry camp • rained some • weather warm[26]

[Jan.] 6) went 4m E • found water & grass [Indian Ranch] plenty of Indians horses have Been here of late • lay in camp the rest of the Day[27]

[Jan.] 7) lay in camp this Day • Killed four oxen that had given out for beef • we are in a narrow vally a consi[d]erable [Young didn't finish his thought] • grass some salt water—some brush for wood

[Jan.] 8) went 8m • campt on a spring [Post Office Spring] good grass • SE

[Jan.] 9) lay in Camp • had explorers out • brought in favorable news • part of our company shouldering their packs[28]

John Brier gives more details about the Indian village:

[A] collection of willow-woven, thatch-covered huts had a cheerful outside for eyes long accustomed to look upon an uninhabited waste. With a single exception, the lodges had been vacated in haste. An old squaw alone remained who, from the doorway of her hut, scolded the intruders with a vehemence that could not be misunderstood. Earthenware, baskets, bridles and hair ropes were much in evidence, while great heaps of offal and the bones of horses betrayed the preferences and predatory habits of the natives. It was their custom to drive animals from the outlying ranges of Southern California and slaughter them in this desert home. (1911a, 4)

James Brier described the location after he returned to the Panamints in 1873: "You all remember the Muskeet Swamps & Indian Springs & Wickeups You remember the very high Mt. East of this spot. That Mt. is the location of the Celebrated Panamint Mines that were discovered since my trip back as a guide & as a result, there is now a populous mining town at this point & a good stage road to it all done, since I was there" (1876).[29]

Post Office Spring, Panamint Valley

From the Indian village they traveled seven miles south to Post Office Spring at the base of the fanglomerate (welded gravel) cliffs below Middle Park fan on the east side of Panamint Valley (three-quarters of a mile south of today's Ballarat). From here, explorers were sent west to find a way to cross the Slate and Argus Ranges. James Brier recalled: "Capt Town & I explored the West rim of Panamint [Valley] where we found an Indian Trail over a very steep pass which I condemned as impassible. The Captain differed & turned back to camp. I went on south many miles farther & discovered a large canyon [Fish Canyon]. I then returned to camp which I reached about 10 p.m." (1896). Brier's "condemned" pass was in the vicinity of the Trona-Wildrose road (Highway 178) that winds up the rugged merger of the Slate and Argus Ranges.

While James was out scouting, back at Post Office Spring, Townes's mess (Mastin, Crumpton, the Turner brothers, and Townes), the last of the Bug Smashers, decided to strike out on their own. Juliet remembered that Mastin and Crumpton "owned the only flour we had, so they baked up their dough, except a small piece, which I made into twenty-two little crackers and put away for an emergency" (1898). James said, "Capt Haynes took out a five dollar gold piece & offered it to the men for one biscuit but they refused it. The Captain wept & said to me 'I have the best 160 acres in Knox Co. Ill. 100 stock hogs, & 2000 bushels of old corn in crib & here I cannot get one biscuit for love or money.'" The boy, John Colton, who had been collecting firewood hoping for a small biscuit, was also unrewarded and "looked sad & disappointed & went away to his blankets" (1896).

Ed Bartholomew remembered, "You, [Asa Haynes, were] the Captain of our little band of lost Jayhawkers. Many are the times I have thought of you, feeble & old as you were then [age forty-six] & starving for want of some nurrishing food. I can see in your hand that five Dollar gold piece you held out so pittiful for that little hard dried up buisket not two inches in diameter" (1885).

James Brier recounted this part of their trek in a letter to the 1896 Jayhawker reunion:

> Next morning the Jayhawks packed up & started for the condemned pass [between the Slate and Argus Ranges], but we remained [at Post Office Spring] until noon to assist our two [five] comrads to pack up—not one word was spoken all that forenoon (all too full). When ready, they turned their faces away & reached out their hands. Not a word was said. As they receded, my little group stood with eyes dimmed with tears & bitterly thought of the morrow. Well we gathered up our oxen & packed up & wearily followed on. At sunset we came to a part of the Jay Hawks & other stragglers who were waiting for us. These were Carter, John Groscup, Harry Vance, M. Gould, Father Fish, & Isham. Well we slept at that point.[30]

Thus, the northern contingent (minus the Bug Smashers) crossed Panamint Valley. Some camped at the small hills three miles northwest of Water Canyon. Several men decided Brier's route farther down the west side of Panamint Valley was better than the "condemned pass," so they joined the family. This ad hoc group traveled south another twelve miles, following the route Brier had scouted a couple days before to today's Fish Canyon that cuts deep into the Slate Range. James Brier recalled:

Next morning I started to explore the big Canyon [Fish Canyon] some 12 or 15 miles away leaving the rest to follow. Mrs B——— was taken that morning with [a] sick headache & spent a most suffering day. I reached the mouth of the Canyon about 2 PM went up 2 or 3 miles untill it closed up to 20 feet in width, with walls on either side overhanging or perpendicular—*A Silent Sepulcre*. Here I found a little damp sand, & scooping out a hole a little water arose. Slakeing my thirst I went on 2 or 3 hundred yards when I found the Canyon walled in but the kind wind of long ages had blown sand enough from the hills above, at a low point of some 50 feet, to make a windrow of sand to the bottom of the pass. With my hands I tore off the apex of this windrow & made a way out on to good ground. I reached the top of the pass [Manly Pass, Slate Range] & seeing our way clear I returned & met the Company at the mouth of the Canyon about sunset. We reached the sepe of water about dark driving our poor famished oxen up the Canyon above us & then we spread our blankets on the sand in this silent & desolate place, & gave ourselves over in the hands of the great God for protection. The place was not only dismal in the extreme, but dangerous. Indians were watching us from the heights of that I was aware, because I saw there fresh tracks in the wet sand in the Canyon in the morning. They could have rolled rocks on us from above & have buried us. I was aware of the danger but said nothing.

In the morning we found that about one half of our oxen had quietly passed us & as we supposed, had returned to the last water. (1896)

Brier's "Silent Sepulcre" was in a dogleg of Fish Canyon at the foot of a tall, dry fall.

Juliet said, "By digging the company managed to scoop up about a pint an hour. Coffee and dried beef kept us alive till morning, but the moaning of the suffering cattle was pitiful" (1898). James had found a sand-filled *tinaja* (tank) where the only way to extract water was to dig a hole and scoop out the water as it seeped into the hole. The dry sand at the surface kept water from evaporating, and the impervious rock below prevented water from percolating into the aquifer. Once the trapped water was depleted, it could be recharged only by another hard rain. Thus, when Manly and Rogers dropped into Fish Canyon about a week later, they found the tracks of the Brier party who "had camped here and had dug holes in the sand in search of water, but had found none." The Brier bunch had found water but had depleted the *tinaja*. Thus, Manly and Rogers "staid all night here and dug around in some other places...but we got no water anywhere" (Manly 1894, 155; see also L. and J. Johnson 1987, 74).

Our family also found water at the base of this high, dry fall in Fish Canyon. With a possible reward of fifty cents each (back when fifty cents bought something), Eric and Mark, our young sons, dug a hole with two small juice cans in what appeared to be dry sand. About two and a half feet down they struck damp sand, and after digging a little deeper, potable water seeped into the hole. (Yes, they got the reward.) While our boys were digging, LeRoy and I followed an old trail to the summit of Manly Pass, looking for Father Fish's bones (none found). At one spot we noted an Indian rock alignment (geoglyph) whose meaning is now lost. The Brier party may have followed the same Indian trail.

Brier Route over the Slate Range

James Brier continued his story of their Slate Range crossing:

> Well I started on with my family, Groscup & Isham. With much effort we got out of the Canyon on my windrow of sand [and] in an hour we were on top of the pass.
>
> After going down the western slope in another Canyon [middle fork of Isham Canyon] about 300 yards we suddenly came to a jumping off place of about 6 feet. Groscup & I built a stone bridge down to the level over which all passed in safety. Groscup then took the lead with his one ox, & mine followed. He soon found other jump offs but the madcap did not wait for a Bridge but forced his Ox to jump & mine followed suit. So on we went until about 3 PM we found ourselves faceing what is now known as Borax Lake [Searles Lake] & right East of your camp.[31] (1896)

They descended the range through the middle fork of Isham Canyon, the only canyon branch obvious from the pass. (In the upper part of the main fork is a twenty-foot fall that would be very difficult to get oxen around.)

In Searles Valley the Briers came upon two Germans camped by a rainwater puddle—probably John Galler (Goller) and his friend John Graff. Juliet Brier remembered:

> We came upon two Germans of the company, who had gone ahead. They were cooking at a tiny fire.
>
> "Any water?" asked my husband.
>
> "There's vasser," one said, pointing to a muddy puddle.
>
> The cattle rushed into it, churning up mud, but we scooped it up and greedily gulped it down our burning, swollen throats. Then I boiled coffee and found the pot half full of mud, so you can see what the water was like. It was awful stuff, but it saved our lives. (1898)[32]

Earlier in her story Juliet said, "Our coffee was a wonderful help and had that given out I know we would have died."

In the meantime, on the west side of Panamint Valley, the remainder of the northern contingent (most of the Jayhawkers and other men) entered Water Canyon a short way and ascended the juncture of the Argus and Slate Ranges through a side gulch west of the Trona-Wildrose Road (Brier's "condemned pass"). In the northern reach of Searles Valley they found a skim of water on a small playa just west of today's road. The mountains funneled the men due south down the valley toward Searles Lake. Sheldon Young related in his diary:

> [Jan.] 10) started again on our way for the west [from Post Office Spring] · went 11 m SW crossing a range of mountains [between the Slate and Argus ranges west of the current county road] · campt on the side of the mountain · Dry Camp
>
> [Jan.] 11) this Day went over a range of mountains verry steep & Difficult · found standing water at the foot · in this valley is a lake [Searles Lake far to the south] · went 14m SW · had a Dry camp
>
> [Jan.] 12) went 12m SW · came to the lake · proves to be salt [and borax] · no fresh water · Dry camp · Dull prospects ahead

Inexplicably Young gives an additional twelve miles of travel on January 12 when everyone else was either searching for water or waiting at Borax Flat for the searchers to return. His previous two mileages (eleven and fourteen) place him at or near today's Searles Lake (or Borax Flat). Since the landmarks of Indian Ranch, Post Office Spring, and "Borax Lake" are still there and corroborated by several other '49ers, mileages and location are easy to determine. The additional twelve miles is an authenticable error—a warning to researchers who use diary mileages: just because it's in writing does not mean it is right. That is why description in a diary and corroboration are so important. Often a diarist is interrupted while writing and later repeats a mileage already given, or entries listed as written daily are actually collectively written when the diarist has a free moment—another opportunity for mistakes to occur.

At today's Searles Lake, James Brier said: "To our horror, the water was impossible to drink. It was what is now known as Borax Lake. Here we overtook the main body of the Jayhawkers. Next day we sat down to die" (1886).

CHAPTER 8

SEARLES VALLEY TO SOLEDAD PASS

If we had a hope—it was only to have food and drink
to satisfy the terrible cravings of nature.
That one thought swallowed up all others.

—Charles Mecum, 1872

Tramp, tramp, tramp, through the weary desert sands.
Often on the point of starvation and in some cases starvation indeed,
to say nothing of the agonies of thirst which no pen can describe,
such were our hardships.

—Dow Stephens, 1893

Searles Valley

Years later, in 1896, James Brier described what happened in today's Fish Canyon in the Slate Range after he left with his family and Groscup and Isham as he pushed ahead of the others:

> The oxen of Carter, Vance, Gould, Fish with 2 of mine were missing. Carter went back for the runaway oxen. I left Father Fish in the care of Harry Vance, for he [Fish] was too weak to travel.... Carter found all the oxen near the mouth of the Canyon. They packed up by noon & started. Father Fish took hold on the tail of his ox & was helped up the windrow out of the dismal Canyon. He held on to the tail 3 or 4 hundred yards when he reeled & fell to rise no more. The boys were so excited that they forgot to leave him his blankets. I sent back for him the next day but he was found dead. The story of Isham's death you all remember.

Brier wrote to a reunion in 1876: "Old Father Fish fell to rise no more on the Mt this side [of] Panamint Valley, that is, west of the valley & Isham his driver was buried 3 or 4 miles south [southeast] of Providence Spring. The Jayhawks voted me his watch & pocket book. His Brother came on from Michigan in '51 &

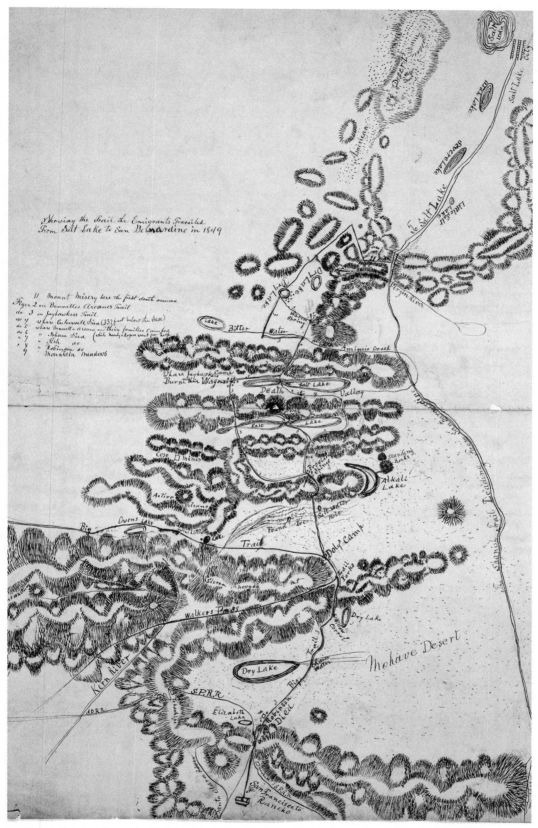

FIGURE 8.1. Routes drawn by William Lewis Manly sent to T. S. Palmer about 1894. Salt Lake, Utah, to Rancho San Francisco, Santa Clarita Valley, southern California. Detail, HM 50895, Palmer Collection. Courtesy of the Huntington Library, San Marino, CA. Discussion of map is in Appendix F.

found me in Marysville; to him I delivered the property." In 1886 James added: "Next day [after finding Providence Spring] we sent two men back for Isham. They found he had crawled 4 miles on his hands and knees toward our camp-fire, and had fallen over and died. They dug a grave and buried him."

Charles Mecum refers to the day the Jayhawkers went back with water to try to bring the missing men into camp: "We left our old friend Fish on the top of the mountain and Isham in the afternoon of the same day. I have always remembered that to leave those two men under the circumstances we were obliged to, were the most trying moments of our trip" (1872).

The Jayhawkers told Manly, "Some took two canteens of water and hurried back to Mr. Ischam, whom they found still alive but his mouth and throat so dry and parched, and his strength so small that he was unable to swallow a single drop, and while they waited he breathed his last. With their hands and feet they dug away the sand for a shallow grave, placed the body in it, covered it with his blankets, and then scraped the sand back over again to make a little mound over their dead comrade" (1894, 340).

About a week later, Manly and Rogers followed the Brier group's tracks out of Brier's "Silent Sepulcre" in Fish Canyon and "found the body of Mr. Fish with some sage brush thrown over him." His body was "laying in the hot sun, as there was no material near here with which his friends could cover the remains" (1894, 155; L. and J. Johnson 1987, 74). When Manly and Rogers returned with supplies to rescue the '49ers remaining in Death Valley a month later, they turned aside from their route to pay their respects to Isham:

> There was now seven or eight miles of clean loose sand to go over [from Providence Spring], across a little valley which came to an end about ten miles north of us, and extended south to the lake where we went for water on our outward journey and found it red alkali [Searles Lake]. Near the Eastern edge of the valley we turned aside to visit the grave of Mr. Isham, which they had told us of. They had covered his remains with their hands as best they could, piling up a little mound of sand over it.
>
> Our next camp was to be on the summit of the [Slate] range just before us, and we passed the dead body of Mr. Fish, we had seen before, and go on a little to a level sandy spot in the ravine just large enough to sleep on. This whole range is a black mass, rocky piece of earth, so barren that not a spear of grass can grow, and not a drop of water in any place. (1894, 190)

Today Fish and Isham are commemorated with the name of each on a canyon, one on the east and one on the west side of Manly Pass in the Slate Range.

Even with all their pain and distress, Edward Doty remembered the spark that kept the Jayhawkers moving and ultimately saved their lives: "i often think of my old companions and our troubles & in imagination can see them marchin along with packs on our backs & sitting around the camp fires discussing the probabilities ahead & i always think with pleasure of the harmony that existed among the Jayhawkers in all their troubles & ther good will towards each other" (1876).

"Terrible cravings, agonies of thirst," "swollen tongues and splitting [lips]," and "the awful silence" were some of the vivid descriptions Jayhawkers used to describe the torture they endured as they trudged the next 175 miles west and then south from Death Valley. Juliet Brier remembered:

> Our greatest suffering for water was near Borax Lake. We were forty-eight hours without a drop.... It was always the same—hunger and thirst and an awful silence.... My husband tied little Kirke to his back and staggered ahead. The child would murmur occasionally. "Oh, father, where's the water?" His pitiful, delirious wails were worse to bear than the killing thirst. It was terrible. I seem to see it all over again. I staggered and struggled wearily behind with our other two boys and the oxen. The little fellows bore up bravely and hardly complained, though they could hardly talk, so dry and swollen were their lips and tongues. John would try to cheer up his brother Kirke by telling him of the wonderful water we would find and all the good things we could get to eat. Every step I expected to sink down and die. I could hardly see. (1898; see also Belden 1954, 26)

Near Borax Lake (today's Searles Lake), James Brier's party caught up with the Jayhawkers.

"If You're Going Through Hell, Keep Going"

As far as we know, all the men went to Searles Lake hoping its water would be fresh, for, as Manly said a week later when he and Rogers passed that way, "on its opposite side, perhaps two or three miles further, we believed we could see trees; and, if so, the lake must be fresh." But such was not the case. To the Jayhawkers' great despair, "the water was impossible to drink." They camped on the sand not far from the wide swath of mud now called Borax Flat. "Before we reached the lake, we came to soft mud where the water had receded. We had half a mile yet to go before we reached the water we so much desired. We waded on as fast as we could through the mud, but alas, when we got to the water to our disappointment it was a strong alkali, red as wine and slippery to the touch" (L. and J. Johnson 1987, 76).

The wished-for trees were the conical-shaped Trona Pinnacles Manly called "Standing Rocks" on a map he sent T. S. Palmer. The wine-colored water was caused by halophilic bacteria found in brackish lakes throughout the world (Map 1890; Manly August 1890; see also L. and J. Johnson 1987, 22; and Post 1977, 150–53).

In the spring of 1873, James Brier guided a group of miners to prospect the Panamint Range and described the lake: "You remember that after leaving Panamint Valley, you crossed a rough mountain & came to a Salt Lake, where I & my family, with Carter & Groscup found you at mid night—but was not that a time! Well that lake is not salt as we supposed, but is Borax. Here are the great Borax Mines or washings. That Glorious Spring brook, discovered on the next morning by Deacon Richards, is now called 'Providence Spring' on account of the many lives it has saved" (1879; also in L. and J. Johnson 1987, 180). The brine from Searles Lake is still mined for boron, sodium, potassium, and related minerals.

While scouts searched for potable water, "Mrs. Brier retired to the shadow of a great rock to pray" (John Brier 1911a, 4). At Pioneer Point, about a mile north of today's town of Trona, a granite formation, Point of Rocks, juts into the valley. As the only woman in this barren valley, Juliet sought some privacy among the rocks—to pray and to rest in a spot of shade from the unrelenting sun.

This was another low point in the Jayhawkers' awful trek. As Harrison Frans wrote to the 1891 reunion, "I say Histry does not tell of it. unfed, unarmed, un-clothed in an unexplored Wilderness where civilised man never trod before and we all truged along the beste we could without much rangling, seamingly happy tho we did not whistle, neither did we Sing. and I will tell you why, because our toungs and lips were so parched for the want of water, we could not."

Dow Stephens also recalled the horrors: "Those desert times, starvation times, hard times, good times, times of gold, times never to be forgotten. Times, too, that life was very uncertain from one day to the next, when our comrades could go no farther, and fell by the wayside never to rise again. Then we would wonder whose turn would come next" (1902).

The Jayhawkers and Brier family were on the verge of thirsting to death when they sent out scouts to look for water from their camps beside the Borax mud-flat north of Searles Lake. Charles Mecum: "That verry night after crossing that desert came the cheering words from that good woman Mrs. Brier: the darkest time is just before day and other similar expressions of cheer. the next morning Doty, Rude and Deacon started in persuit of water. all returned but Deacon with long faces and bad news but when he came he brought the glad tidings of plenty of water in a canion 3 or 4 miles away" (1872). John Brier:

> While she [Juliet] was yet declaring her certainty that deliverance would come, "Deacon" Richards bounded into the circle, shouting, "Water! Water! I have found water!" Four miles away, hidden close in at the base of

the mountain, was a clear brook that ran for a space and sank out of view in the desert sand. The heroic discoverer had not been willing to spare time for the slaking of his own thirst.... There was bustle in that camp, and within the hour men who had resigned themselves to die found a new joy in living. (1911a, 4)

In 1896 James Brier wrote, "That Spring Branch is now called Providence Spring on account of the number of lives saved by it since our day. From there to the Lake [is] 4 miles south.... At Providence Spring I found a Jayhawke ox shoe which I brought home as a memento [on his return trip leading some miners]."

Providence Spring Branch

We have searched for years to find a reference or a map (other than the '49ers') to nail down where Providence Spring is today, but as yet we have found no reference to the name. Historians and trail buffs have speculated that it is located along the eastern base of the Argus Range somewhere north of Trona, even as far as Bruce Canyon and Carricut Lake. Forty-niner mileages vary concerning their camps on Borax Flat to the Providence Spring branch. Mecum (1872): "plenty of water in a canion 3 or 4 miles away." Young (January 13, 1850): "this Day went 5m N to a spring in a Deep rocky Kenyon."[1] Asa Haynes (1849–50): "north 6 miles. Water." James Brier: "at the foot of the high peak [Argus Peak]" (1876) and "up a Canyon" (1898). Manly (1894, 189–90): "a faint running stream which came out of a rocky ravine and sank almost immediately in the dry sand," and "a small weak stream, but very good, that came from the rocky cliffs—but there was no feed for the cattle, except now and then a bunch of sage" (see also L. and J. Johnson 1995). Manly drew two maps, each with a stream issuing from "Providence Springs." Coupling these maps to the rest of the data, our conclusion is the argonauts camped on the stream that issues from Indian Joe Spring (James Brier 1879; John Brier 1911a; Manly 1894, 189; Map 1890; Manly ca. 1889 and August 1894; Manly in L. and J. Johnson 1987, 122).[2] Juliet Brier recalled the lifesaving water: "I have always believed Providence placed it there to save us, for it was in such an unlikely place" (1898). Manly to the Jayhawkers: "At the spring you found after *Isham* died the rocks you put your camp kettle on still stood as you left them" (March 16, 1890). Sheldon Young recorded their finding Providence Spring:

> [Jan.] 13) this Day went 5m N to a spring in a Deep rocky Kenyon • the Best water we have found in some time • not much grass some Brouse • Killed a cow • some of our men not in camp yet • two we Do no[t] think can be got into camp • they were so fatigued out that they cannot come in

[Jan.] 14) this [day] went out in search of the two men · found them both
    Dead · their names were Fish from Indiana & Icem [Isham] from
    Michigan · Both died of fatigue · lay in camp this Day · Killed 6 oxen
    & Dri[e]d [meat] for the Journey · Six men came up to camp this eve-
    ning from four waggons that were behind packing through
[Jan.] 15) this Day lay in camp · nothing occured worthy of note

Some of the "six men" Young mentions were teamsters from the Bennett-
Arcan family wagons still in Death Valley—Samuel Abbott (a Bennett teamster),
Harry Vance, a man named Harrison, and one (or maybe two) Achisons.[3] Juliet
said two men, Patrick and Loomis St. John, "came to us at Providence Springs in
a state of starvation [and] asked help. Mr. B. told them they would be welcome
to such as we had—meat and coffee and asked them to assist me with the oxen
which they were most willing to do" (1905).[4]

Over the Argus Range into Still Another Desert (Indian Wells Valley)

Many of these men were now without boots or shoes, which had long since worn
out. They wore hats to shade their eyes, and those with access to needle and
thread had patched their travel-worn clothes. They were bearded, not only to
protect their faces from the relentless desert sun, but because a razor was unnecessary weight. They carried their few worldly possessions on their backs or tied
in bundles on the backs of their emaciated oxen. Juliet Brier, the only woman
in the northern contingent, said, "We looked more like skeletons than human
beings. Our clothes hung in tatters. My dress was in ribbons, and my shoes hard,
baked, broken pieces of leather. Some of the company still had the remains of
worn-out shoes with their feet sticking through, and some wore pieces of ox hide
tied about their feet. My boys wore ox hide moccasins" (1898).

Apparently, at Providence Spring the Jayhawkers made the momentous decision to divide into smaller units with the hope of finding water and grass enough
for their needs. When Manly and Rogers caught up with the Doty division at
Indian Wells at the base of the Sierra Nevada (it included Tom Shannon, Dow
Stephens, probably Leander Woolsey, and others), the Briers were a day behind
them and the Asa Haynes division a day ahead (Manly 1888, 126).[5] "Men from the
various parties scattered all around the country, each one seeking out the path
which seemed to suit best his tender feet or present fancy, steering west as well
as mountains and cañon would permit, some farther north, some farther south
and generally demoralized" (Manly 1894, 338).

From Providence Spring the various groups entered East Wilson Canyon to
ascend the Argus Range. Here Juliet Brier had an alarming fright. Her son John
explained, "The next few days were among the worst of the journey, and we were

in the poorest condition to endure them. We went along an Indian trail into a defile, in one of the branches of which my older brother [Christopher, then age nine, herding some oxen] lost his way, and my mother was nearly distracted with fear of his capture by Indians before he again joined us" (1903, 456).

After about ten miles of loose sand they camped in Sweetwater Wash, on the west side of the pass that separates East and West Wilson Canyons (an ironic name—*no* water, neither sweet nor salty). The Jayhawkers had a dry camp and another view of the "verry lofty" Sierra Nevada—"father of all the mountains we had ever seen" (Manly in L. and J. Johnson 1987, 126). Their goal—Walker Pass—was about fifty miles away, almost in sight but around a spur of the snowy southern Sierra Nevada.

When Manly and Rogers brought the Bennett and Arcan families out of Death Valley about a month later, Manly described the trail after they left Providence Spring.

> As soon as ready we started up the [Wilson] cañon, following the trail made by the Jayhawkers who had proceeded us, and by night had reached the summit, but passed beyond, a short distance down the western slope, where we camped in a valley [Sweetwater Wash] that gave us good large sage brush for our fires, and quite a range for the oxen without their getting out of sight. This being at quite a high elevation we could see the foot as well as the top, of the great snow mountain [Sierra Nevada], and had a general good view of the country. (1894, 227)

In 1873 James Brier retraced his 1850 route (in reverse) and provided mileages, locations, and the best description (other than Manly's) of this part of the trail:

> Let me take you all back. From Indian [Wells] Springs [base of Sierra Nevada] to the 1st range of Owen's Mt [Argus Range] is 25 miles. We crossed this Desert [Indian Wells Valley] in [18]50 you all remember, in a little less than 2 days. From the spring [pond at Paxton Ranch site] at the mouth of the Canon [Deadman Canyon] that we came down to the top of the Mt [Argus Range] where we camped is 10 miles. from the summit camp [Sweetwater Wash] to the spring branch found by Deacon Richards at the foot of the high peak [Argus Peak] is 10 miles. That Spring Branch is now called Providence Spring [flowing from today's Indian Joe Spring] on account of the number of lives saved by it since our day. From there to the lake [is] 4 miles south. The Lake [Searles Lake] was not salt but Borax. It is now famous for that Mineral. From the Lake to the spring in Panamint Valley [Post Office Spring] where Capt. Town & Co left me & I joined the Jayhawkers is 25 miles. (1876)

From Sweetwater Wash, two routes down the western slope of the Argus Range lay before the Jayhawkers—West Wilson and Deadman Canyons. Deadman Canyon debouches a couple miles northeast of the salty pond at Paxton Ranch site in Indian Wells Valley, and Wilson Canyon debouches about four miles farther north.

Through Indian Wells Valley

It is important to point out here that the miles Sheldon Young (and probably Asa Haynes) tramped through loose sand tend to be about 10 percent larger than what one might calculate by drawing a mileage path on a map. The sinuous course of such a foot trail around sage and bitter brush increases the miles traveled, as does the perception of distance from the added effort of walking in sand in a starving and dehydrated condition. Many early travelers also speak of deceptive distance in the desert.

Most readers cannot imagine how difficult it is to walk in loose sand until you have done so for hours. It requires a prodigious amount of labor with many more muscles than walking on a firm surface. When Dick Bush and LeRoy followed Manly and Rogers's route from Death Valley to the casa site of the Rancho San Francisco, they sometimes walked backward in the sand in Indian Wells Valley to ease the strain on their aching legs. Manly said of Mrs. Bennett and Mrs. Arcan, "Their work in walking was almost more than they could endure,... and [they] could hardly stand upon their sore feet and tired limbs." "One of the women held up her foot and the sole was bare and blistered. She said they ached like toothache" (Manly in L. and J. Johnson 1987, 124; see also Manly 1894, 234).

Sheldon Young provides an overview of the trek from Providence Spring (Indian Joe Spring branch) in Searles Valley, across the Argus Range and Indian Wells Valley, to Indian Wells at the base of the Sierra Nevada:

> [Jan.] 16) this Day crossed Owens mountain [Argus Range] · went 12m W · crossed in a Deep Kenyon · had a Dry camp near the top in Sight of the Serenevada [Sweetwater Wash] · th[e]y look verry lofty
>
> [Jan.] 17) had a frost last night · the first we have had since the 10[th of] Dec · went 10m SW into a vally of salt & alkali [west Wilson Canyon into Indian Wells Valley] · found some grass & campt
>
> [Jan.] 18) this Day went 10m SW · left old Brin [a brindle ox] · found fresh water [China Lake] some grass · Saw Signs of game some sheep tracks
>
> [Jan.] 19) went 17m SW · canpt [camped] on a spring [Indian Wells] · grass water no wood cold frosty nights warm Days · we are now near the [Walker] pass on elevated ground

Young's mileages indicate he entered Indian Wells Valley via West Wilson Canyon. He mentions "some grass" but does not mention water, possibly because the pond at Paxton Ranch site was barely potable. His mileages indicate he (and comrades) went southward to China Lake, where they "found fresh water."

China Lake is another playa with only a couple permanent places where potable water comes to the surface. But a skim of water from recent rains probably covered part of the playa where the men could dig holes large enough to dip out water. These same rains left puddles in small depressions elsewhere in Indian Wells Valley. Thin sheets of ice covering some of these depressions saved the lives of Manly and Rogers a day or two after Sheldon Young crossed the valley. Surface water on these puddles froze at night, but water underneath percolated into the ground, leaving a suspended sheet of ice.

From China Lake, Young likely followed an Indian trail across Indian Wells Valley to fresh water at today's Indian Wells at the foot of the Sierra Nevada.

The Doty contingent of Jayhawkers first stopped (or paused) at Paxton Ranch site, used the salty water there, and rather than continuing south to China Lake they crossed the valley directly to Indian Wells (as did the Brier family a day later). Ed Doty's group explained their route to Manly and Rogers, who later used it (in reverse) when they returned to Death Valley to escort the Bennett and Arcan families out of the desert. At the Paxton Ranch site with the families, Manly said: "Camp was made at the salt water hole, and our wheat and meat boiled in it did not soften and get tender as it did in fresh water. There was plenty of salt grass above [about?]; but the oxen did not eat it any more than the horses did, and wandered around cropping a bite of the bitter brush once in awhile, and looking very sorry. This was near the place where Rogers and I found the piece of ice which saved our lives" (1894, 228).[6]

Asa Haynes's contingent may also have turned south to China Lake, as apparently Sheldon Young's mess did. Haynes's diary says, "thence South 8 mile. Water and grass"; Young says, "10m." The discrepancies between Haynes's and Young's mileages are frustrating. This is another case where "the clues that remain will always prove insufficient to our curiosity" and "[when] different parts of the evidence contradict each other; using your own judgment about it all means that you must face such contradictions squarely" (Caesar 2014, 34–35; Marius 1999, 26, 48). For this book I choose to follow Sheldon Young's diary rather than trying to correlate, coordinate, or compare the differences between his and Haynes's diaries.

Oxen

A word of compassion is due for the exhausted, hard-used oxen that were in as poor a condition as their human masters. They had pulled heavy wagons for more than two thousand miles, much of it over roadless country—through sand and over mountains. The deserts offered them poor feed and barely drinkable water. Their metal shoes had worn through, and the lucky ones wore moccasins made from the hide of a brother ox slaughtered to feed their masters. These impoverished oxen were the walking victuals for the humans. William Lorton had said months before, "to kill our cattle is to take our life" (May 29, 1849). Now the saying was reversed: to kill our cattle is to save our life. One by one, the weakest ox was put out of its misery and eaten—"Old Ox hide and young Ox hide, Boiled, Stewed & Broiled…Ox Blood hot and Ox blood cold, Ox blood thick and Ox blood thin" (Stephens 1887). "When we cut them up, we found their bones had no marrow in—nothing but blood and water. Their flesh was poison, covered with yellow slime. We lived on the hides, boiled" (James Brier 1886). Juliet added: "My husband and I and the poor children and St. John and Patrick lived on coffee and jerked beef, except when we killed an ox for a new supply. Even then there was not an ounce of fat in one and the marrow in their bones had turned to blood and water" (1898). Manly commented on the plight of the oxen in various accounts: "One cannot form an idea how poor an ox will get when nearly starved so long. Months had passed since they had eaten a stomachful of good nutritious food. The animals walked slowly with heads down nearly tripping themselves up with their long, swinging legs. The skin loosely covered the bones, but all the flesh and muscles had shrunk down to the smallest space" (1894, 242). In an 1891 letter he added, "Poor ox meat but little better tharn basswood bark strong scented with sage, & the marrow in their bones run & looked like corruption.... [N]o ox was ever allowed to die & be left" (1891). "Our feet needed new protection as well as those of the oxen. We dried the meat, made soup of the bones, and a sort of pudding out of the blood" (Manly in L. and J. Johnson 1987, 128).

Their bones were so prominent, Harrison Frans said, an "ox if turned back down would then have made a better harrow than an Ox" (1889). Then he summed up their trials: "[When] I look back and Scan over the rout we traveled, it makes me Shudder. I often think of how we Suffered and I burst in[to] tears and cry a loud. as I grow older I more than realise our condition" (January 26, 1892).

If you're going through hell, keep going.

Indian Wells, Base of the Sierra Nevada

James Brier: "You remember, at the foot of the Sierra Nevada, some Indian Springs, walled up with rock. well, at this point stands, now, a Hotel. I rode up to the door, & told the Land Lord that I held a claim on the ground, for I had slept there 25 years ago with my family. He looked stunned & could scarcely believe me" (1879). Dow Stephens recalled:

> In 1864 I was traveling down on the eastern side of the Sierra Nevada mountains, via Owens' lake. When at Indian Wells, as it is now called, (there is a [stage] station there of that name), I fell in company with a

man that had a quartz mill and mine out near Death valley. He told me that some of his men had found the skeletons of nine persons all together, and a little wall of rock built around them. That, I suppose, is nine of the eleven that were lost in Death valley, that started to pack through on foot [now called the Pinney-Savage group]. I believe though, as near as I remember now, there was sixteen in the party [Bug Smashers] that left us. He also stated that they had found the bones of two others, further south. That, I suppose, was old Mr. Isham, the fiddler, and another man whose name I have forgotten [Mr. Fish], who were with our party. You, of course, all remember them. I requested him to take the remains of all he could find and give them a burial, which he promised he would do. (1880)[7]

At Indian Wells five miles north of the entrance to Walker Pass, "Capt. Doty and his mess had stopped here to dry the meat of an ox that had given out, while the others [with Asa Haynes] had gone ahead and left them. They staid two days, and while waiting, Brier and family came up and all went on together" (Manly, January 1889, 174).[8] A few months after Doty's death in 1892, Dow Stephens wrote: "You all know that Capt Edward Doty was a first class man in every respect and there was none that knew it much better than I did, being in the Same Mess" (1892).

Manly and Rogers sneaked up on the Doty contingent at Indian Wells, greatly relieved to find they were not Indians. Stephens remembered, "They overtook us at what is now called Indian Wells. Stayed all night with us and then went ahead" (1884). Manly relates what the Jayhawkers told them.

Mr. Brier and his family were still on behind, and alone. Every one must look out for himself here.... These men were not as cheerful as they used to be.... An ox was frequently killed, they said, and no part of it was wasted. At a camp where there was no water for stewing, a piece of hide would be prepared for eating by singeing off the hair and then roasting in the fire. The small intestines were drawn through the fingers to clean them, and these when roasted made very fair food.

They said they had been without water for four or five days at a time and came near starving to death, for it was impossible to swallow food when one became so thirsty. They described the pangs of hunger as something terrible and not to be described. (1894, 161–62)

Manly provided additional information in his 1888 account:

As some were sitting up tending the drying of meat and keeping up a fire, we sat up quite late talking with them, although we had not slept a wink the night before. They told us of the death of Mr. Fish and Isham, and the hardships and suffering they had had.... They then spoke of the big snow mountain beside us [Sierra Nevada] and said if their big trail led into it they would have to leave it [because of snow]....

In the morning we shouldered our knapsacks and took the trail which followed along the foot of the mountain for seven or eight miles; here a small branch turned north [northwest to Walker Pass] and seemed to go into the snow, but the broad one bore south and [then southeast] away from the mountain and across a level plain to a quite low range [El Paso Mountains]. (L. and J. Johnson 1987, 84)

"We all, indeed, were getting worn down in body as well as feet," Manly remembered (ibid., 129). Dow Stephens concurred: "Weary, faint, and almost dying from want of food and water; how sad to think of the past, and to see in the midst of that lonely Desert, our comrades dying, and we unable to lend a helping hand. It is still a painful topic for me to talk on, and I seldom speak of it even to most intimate friends.... [T]here are few who can tell the almost incredulous stories that our noble little band of Jayhawkers can relate of their experiences on the Desert" (1894).

James Brier: "Our sufferings were fearful. Our parched tongues protruded from our mouths and we could not eat our jerked meat" (1886).

Stephens:

Speaking of thirst, there is no punishment that has any comparison. It is the most agonizing suffering possible and the feeling is indescribable. Our tongues would be swollen, our lips crack, and a crust would form on our tongue and roof of mouth that could not be removed. The body seemed to be dried through and through, and there wouldn't be a drop of moisture in the mouth. at these times we thought of Barney Ward, and I just can't imagine what we would have done to him had we had him near....

My thighs were not larger than my arm, and the knee joints were like knots on a limb, and on my hip bone the skin was callused as thick as sole leather and just as hard, caused by lying on the hard ground and rocks. We were all of us in a pitiful condition. (1916, 21, 25)

If you're going through hell, keep going.

Walker Pass

The Jayhawkers followed the trail southward from Indian Wells to Coyote Holes at the entrance to Walker Pass (later known as Freeman's stage stop). All that remained of the stage stop in 1964 were some concrete fragments and rusting pipes, and those were gone shortly thereafter (Edwards 1964, 31, 37, 38, 42). The

Los Angeles aqueduct west of U.S. Highway 14 closely parallels the old trail from Indian Wells south to Freeman's (one and a half miles southwest of the current junction of Highways 14 and 178).

The Jayhawkers could see an unusual rock formation now called Robbers Roost about two and a half miles southwest of today's junction: "Some standing columns of rock, much reminding one of the great stone chimney of the boiler house at Stanford Jr., University; not quite so trim and regular in exterior appearance, but something in that order. We reckon the only students in the vicinity would be lizards" (Manly 1894, 233). Tiburcio Vasquez used these rhyolite monoliths as a rendezvous site for his raid on Freeman's stage station on February 25, 1874 (Edwards 1964, 17).

Because of snow, Walker Pass was closed to the Jayhawkers! They wearily turned their backs on the snowy pass and followed the wide, ribboned Indian horse-thief trail south, then southeast across the sand and sage of southern Indian Wells Valley. Historic evidence points to the '49ers following the horse-thief trail via Last Chance through the El Paso Mountains. Asa Haynes's diary says after traveling south fifteen miles from their camp "at the foot of the serinavady mountains," they turned "E of S 9 miles" to the rim of Indian Wells Valley and into Last Chance Canyon, following the Indian horse-thief trail.[9]

Frank Latta told us that in the 1940s, the trail going into and out of Last Chance Canyon was visible from the air, but dune buggies and trail bikes have since obliterated most signs of it. LeRoy and Dick Bush found a remnant (and an arrowhead chipping site) in 1973 north of the crest where it drops into Last Chance Canyon.[10]

Sheldon Young's diary gives a synopsis of his route from Indian Wells through the El Paso Mountains to Desert Spring in Fremont Valley (in the northern Mojave Desert). Manly called Desert Spring "willow corral" and Young describes it as "plenty of willows":

> [Jan.] 20) this Day went 16m S • Struck a large trail [Indian horse-thief trail] • plenty of horse tracks • followed it all Day • had a Dry camp • no grass • weather rather cool • freesing nights
>
> [Jan.] 21) went 14m S [southeast] • struck into a Deep Kenyon [Last Chance Canyon, El Paso Mountains] • found water not much grass some Brouse [later site of Cudahy Camp] • plenty of horse tracks & Indians sighns • verry cold nights
>
> [Jan.] 22) went 8m S through a Deep Kenyon • came into a vally [Fremont Valley, northwest wing of Mojave Desert] • an alkali bottom • found plenty grass & water[11]

[Jan.] 23) went 5m S.W • found grass water & plenty of willows [Desert
Spring, Manly's willow corral] • the country begins to have a Different
look • we are now in hopes of soon getting out

Last Chance Canyon and the El Paso Mountains

Freshwater springs in Last Chance Canyon are the main reason the horse-thief
trail traversed it rather than Red Rock Canyon. Sheldon Young's diary also sup-
ports Last Chance Canyon. After he "came into a vally," he "went 5m S.W • found
grass water & plenty of willows." Had he exited Red Rock Canyon (the later stage
route and current course of Highway 14), he would have traveled a little east of
south for only two and a half miles, not five and a half miles southwest to the
willow corral at Desert Spring.

Manly and Rogers followed the tracks of Asa Haynes's forward contingent
along the horse-thief trail:

We followed down a very steep canyon [Last Chance] and when we came
to its bottom, we found a weak little stream of water and here the party
we were following were camped. No grass, and sage brush scarce.... Here
we found Bennett's and Arcane's four teamsters, with nothing to eat of
their own.... When we got ready to start in the morning, many gathered
around us and wished us good luck.... Some gave us their addresses in
Illinois; among them, the oldest man in the camp, Capt. Asa Haynes, requested us when we got to San Francisco to go to the fort and have the
news sent to their families, telling them how they suffered, starved, and
choked to death on the deserts.... Many thought they were doomed to
perish and died alone and go without burial.... We finally bid them one
and all goodbye and went sadly and silently down the canyon. In eight or
ten miles we reached the open plain and our trail seemed to point across
it (Manly in L. and J. Johnson 1987, 85–86)

Manly spoke specifically of the Bennett-Arcan teamsters: "They were already
out of grub and no way to get anymore. When the party killed an ox they had
humbly begged for some of the poorest parts, and thus far were alive. They came
to us and very pitifully told us they were entirely out, and although an ox had
been killed that day they had not been able to get a mouthful. We divided up our
meat and gave them some although we did not know how long it would be before
we would ourselves be in the same situation" (1894, 162).[12]

The Great Mojave Desert (Fremont Valley, Northern Mojave)

The Garlock Fault created the sharp drop from Indian Wells Valley plateau to the Mojave Desert—that great sea of sand that stretched ahead of them to the southern horizon. Bleak as the sight was, Manly and Rogers pressed southward, leaving behind the fragmented cavalcade of the northern contingent. "A 75 mile plain [was] before us.... Here was a partly dry alkali lake [Koehn Dry Lake], and on or near it was some grass and fresh water"; the same view that greeted the Jayhawkers (Manly in L. and J. Johnson 1987, 129–30).

The Jayhawkers found Desert Spring (southeast of today's Cantil), "which was a pretty good camp, with some wood and water, and a little grass for the stock" (ibid., 130), where Young "found grass, water & plenty of willows" (January 23, 1850). Here the Indians had woven a holding pen for their stolen horses, hence Manly's name, "willow corral." This lifesaving water is commemorated with State Historical Monument 476, a quarter mile off Papas Road on private property. It proclaims: "This spring was [on] an old Indian horse thief trail and later (1834) Joe Walker Trail. The famished Manly-Jayhawk Death Valley parties (1849–50) were revived here after coming from Indian Wells through Last Chance Canyon. This was also a station on the Nadeau Borax Freight Road." Manly described the spring further:

> We came to a bunch of willows growing thick and perhaps ten feet high. They were growing on level ground and the grove was perhaps fifty feet or more in diameter. In it we found a good spring of water, but it could not run but a few feet beyond the willows. The willows in the center of the grove had been cut out and woven into those left so as [to] make a small corral.... We now filled ourselves with water as well as our canteens, and hurried on west [south] in the trail. (L. and J. Johnson 1987, 22 ["Palmer map"], 86–87)[13]

South of Desert Spring, the '49ers found bones along the trail that "followed the highest ground on the plain and the traveling was good.... We found that a little off of our high trail the ground was very soft and light as snuff. An animal that might try to walk in it, would sink six inches at every step" (Manly in L. and J. Johnson 1987, 87; also in JCHL Scrapbook 1).[14]

The Jayhawkers anticipated several days' march to reach the distant San Gabriel Mountains to the south. Their suffering from lack of water was again excruciating during the days spent crossing the Mojave Desert.

Visible foes we feared not:—we could meet them like men—but before the invisible foes of hunger, and thirst, we became as helpless as children. How when the nights became cold we made our beds of burnt sage brush:—which made us as restless and active as pop-corn in a hot skillet. Of our weary march day, after day;—week, after week;—month, after month,—on that terrible desert:—of our sufferings from hunger, and thirst:—of our devouring with savage avidity, *choice bits* of our cattle.... How with *sad hearts* we left our comrades on the desert—not knowing how soon our turn would come:—how wearily we plodded on—all hopes of speedy wealth vanished—only dull despair in its place. If we had a hope—it was only to have *food* and *drink* to satisfy the terrible cravings of nature. That *one* thought swallowed up *all* others.

The *world* knows nothing of that terrible journey: but the *memory* of those days,—with their *hopes*, and *joys*,—their *sufferings*, and *despair*,—are indelibly imprinted on the hearts of the few survivors. They form a *bond* of Brotherhood which time cannot efface. (C. Mecum 1878)

Harrison Frans: I "look back to the Jayhawks worne out and allmoste destitute when they approached the Desert with Capt E Doty and Cpt Hains in command. our little party trudged along, and strange to say only lost 4 out of the party. I think it miraculous. I could tell you of sum laughable occurrances whilst on the Desert, but perhaps they might not bee in order here" (1891).

John Colton, the youngest Jayhawker, was interviewed for an article in the *Kansas City Times* in 1888. This excerpt alludes to their distress on the desert: "So on we staggered, hoping every morning to reach fresh water and grass before night, and lying down every evening in despair.... But still we plodded on, a discouraged party of hollow-eyed skeletons. Rifles had been thrown away, for there was no game, and at any rate we had become too weak to carry them. Then there came a day when we looked at each other and wondered whose turn would come next." Harrison Frans remembered seeing "my Brothers all holding their little frying pans to catch the last drop of blood oozing from a poor old ox starved to death" (1889).[15]

After talking with Jayhawkers Dow Stephens and Ed Doty, Manly related, "Stevens himself awoke at dead of night and seemed perishing. He crawled to Doty's bed and begged for a little sup of water, for he felt himself nearly dead. Doty gave him a swallow which was nearly all he had" (February 1889, 190).

Sheldon Young recorded their travel from the "willow corral" (Desert Spring) across Fremont Valley, the northwestern wing of the Mojave Desert:

[Jan.] 24) went 18m SW on a high piece of country [west of California City] • been verry cool th[rough?] the Day • cold & windy from the west

[Jan.] 25) went 24m S had high tableland · campt by some puddle water · grass Scarce · weather continues cool · Storming on the mountains [Tehachapi or San Gabriel Mountains or both]

Into Antelope Valley (Southwestern Mojave Desert)

Fremont Valley is separated from Antelope Valley by an east-west range of hills and buttes culminating at Soledad Mountain, but both are part of the Mojave Desert. Here the '49ers experienced the longest waterless stretch, extending about thirty-eight miles from Desert Spring in Fremont Valley southwest to Willow Springs in Antelope Valley. To compound their difficulties, the herds of stolen horses had created a web of trails as the Indians drove them northward, and they became difficult to follow. Sand blown by cyclonic storms over the Tehachapi Mountains obliterated the loose network of trails. Some of the Jayhawkers separated to try to find it, for, as Manly said, "the mountain before us [San Gabriel Mountains] was covered with snow and it would be almost necessary for us to cross it in a trail" (L. and J. Johnson 1987, 89).

The horse-thief trail led to Willow Springs in Antelope Valley, the only major natural water source in the heart of the southwestern Mojave Desert. It nestles in a fault just west of Willow Springs Butte, about seven air miles west of today's Rosamond. A southwest trajectory from Desert Spring ("willow corral") slices through solitary Soledad Mountain to Willow Springs where California State Historical Landmark 130 commemorates this important watering hole. It was settled by early pioneers and was a stage station prior to 1875 when the Southern Pacific Railroad was built (Dibblee 1963, 147). Some of the Jayhawkers were able to trace parts of the trail as it veered southwest from Fremont Valley to Willow Springs.

Tom Shannon and Leander Woolsey were hiking together, apparently in the vanguard, when Tom described his near-death experience in this part of the desert.

> That morning Woolsey and me left a dry camp with emty canteens to try and find water in the next range ahead [hills and buttes separating Fremont and Antelope Valleys] which was a days journey. from the time we left camp until noon there had not been a word between us. our tonges were not in working order. about noon we were diveying [dividing], I going west and he bearing more to south. about the time we parted a hare jumped in front of me and I shot his head off. Woo[l] sey saw me but kept on his course and I mine. I reached the foot of the mountain [Soledad Mountain or west Rosamond Hills] about sundown. saw a band of wild horses 9 in number, but could not get nearer than

half a mile of them. I ate some of the young grass where they were feeding and then imagined I was nabuchudnazer [king of Babylonia amid a surfeit of plenty]. But my search for water was in vain and being completely exausted, I stood my rifle against a rock and hung my shot pouch and powder horn on the ramrod, and sat down in an [as?] easy posision as posible to die, my eyes gazing wistfully at the snow capped summit of kowhere [nowhere] mountain some 50 miles to the south [San Gabriels]. in that condition I had a remarkable vision. my whole life from my earliest recolections passed in a panoramic view before me. then, I laughed at the vain strugles of my life that I thought was ended, but it was not to be so. I heard a human voice calling me. I could not make answer, nor did wish to, for I could not realise a possibility of any one coming to my aid. it was Woolsey whom I supposed dead. before that time he had found water at the foot of the mountain [Willow Springs Butte (?) or some rainwater holes] some 2 miles South of me. he had found a mudy pool that you will remember, filled his can and then set out to find me and share my hare for supper, as he had neither meat, maches or gun. That is how Tom can still answer—Here. (1894; also in *Beaver Valley Tribune*, February 4, 1894)[16]

Young's diary is the main source of information for finding the Jayhawker route with the caveat that there were surely individual variants. The northern contingent was still split into smaller groups. Fortunately, Sheldon Young's group found Willow Springs after three days without water. He wrote:

[Jan.] 26) went 3m S · Came to a beautiful spring of water & wood [Willow Spring west of Rosamond] laid over · two oxen killed

[Jan.] 27) went 10m S · lost the trail · sent back to the camp to trace it out · came amongst the sage again · had a Dry camp

[Jan.] 28) bore a W of S [E of S]· Struck trail again · went 16m to the mountains · struck into a large Kenyon [San Andreas Fault] · found grass & water [Barrel Springs southeast of Palmdale] · horse & ox bones about here plenty · Wm. Robertson [Robinson] brought into camp · Died in the evening from the effects of Drinken to much cold water after being without part of the Day [17]

[Jan.] 29) lay in camp · Killed an ox & jerkt it · two of our men out hunting · saw fresh signs of wild cattle & saw wild horses

The distance from Willow Spring to Barrel Springs is about twenty-eight miles. Sheldon's twenty-six miles is suitably close. This section of the trail may

FIGURE 8.2. The '49ers found horse bones on the Indian horse-thief trail in Fremont and Antelope Valleys, Mojave Desert, but they left their ox bones, about 2010.

be where Charles Mecum recalled, "I cannot place the time that we traveled all day and then had to retrace our steps to get a drink" (1872).

During this difficult part of their trek, Ed Doty exhorted his demoralized and exhausted comrades: "'Determine to get through with this, no matter what stands in the way. You can endure much more than you think you can'.... Doubtless many were kept up by these words alone, and the example he set before them" (Manly, February 1889).

If you're going through hell, keep going.

Those who still had oxen took turns killing the least able one for food. At one point it was Tom Shannon's turn to kill an ox, "and all those who had cattle took what they wanted of it, and then permitted those who had no cattle to come forward and take the rest" (ibid.) Most of the men had thrown away their weapons since there was no game to kill, Indians were unseen, and guns were an unnecessary weight, but Tom Shannon still had his yellow-stock rifle (more about it in chapter 10). Edward Bartholomew remembered that dinner:

> Me thinks I see us now, camped there in the sand miles from water, dividing up that old ox that gave out & died that evening, burning off the hair & crisping the hide so that we could eat it & preparing some Tripe (by well shaking) & laying [it] on the coals of a sage brush fire, Making our suppers of these two dishes without water. & the night before, saving the leading-lines & every drop of blood from the ox Tom Shannon killed, after he had lain down to die, the puddings & fine dishes fixed up that night,...

The Several breakfasts on Jerk from the flesh of those old dead cattle. "I say boys, *Didn't they tast good?*" And wouldn't the contents of our fathers *Swill-barrel* [have] added lusciously to our meals in those days? (1888)[18]

In 1903 Tom Shannon wrote to the Jayhawker reunion from the Soldiers Home in Los Angeles: "I hope none of you will become so hungry as to raid Mrs. Briers dinner before she herself decides it sufficiently cooked as we all well remember [what] happened on one occasion when liver was not so easily obtained as now." Wherever this was, the Briers and Jayhawkers were camped together, and Juliet was doing the cooking.

The farther the Jayhawkers traveled, the weaker the men and oxen became, worn down by hunger, thirst, pain, and exhaustion. Eighteen-year-old John Colton probably would have succumbed to death in the Mojave Desert had his friends not cared for him: "Brother Colton will well remember being strap[p]ed on an old ox for several days to be kept with his companions" (Frans 1889).

As they approached the Sierra Palona and San Gabriel Mountains, they came across Manly and Rogers's tracks heading toward today's Barrel Springs. A month later, as the Bennett-Arcan families approached these springs, Manly commented on the Jayhawker trail: "As we approached the low foot-hills the trail became better traveled and better to walk in, for the Jayhawkers who had scattered, every one for himself apparently, in crossing the plain, seemed here to have drawn together and their path was quite a beaten one. We saw from this that they followed the tracks made by Rogers and myself as we made our first trip westward in search of bread" (1894, 245).

John Groscup had the most phonetic and imaginative spelling of the '49ers and wrote his recollection of their desert trek in verse: "Weeak and faint & starving to, wee still this Desert must go threw, but Alas some ware left be hind to stay, which cased [caused] still harder Trials by the way. Wh[e]n in the Grave wee lade them by, not noing how soon that wee Might Die. Still on wee went as wee should, to find some water if wee could" (1885).

Dow Stephens:

I cannot realize yet how near to death's door we all were. It seemed that water was always found at the extreme last moment many times. if it had not been found just at the time and we had to travel on to the next, how many of us would have survived? (not one).... [I]t seemed that the last mile could not possibly [have] been extended to another mile. It was nothing but the indomitable sticktoitiveness that ever brought us through the Desert and Death Valley to pasture green and land of plenty, land that flowed not with milk and honey, but with carny con chilly and tortillas and frijoles. (1905)

Stephens also spoke of that all-important emotion—hope:

The boys that marked the path across the great American desert and pushed his way through the red mans country oblivious of all danger, that could make his supper on a small piece of rawhide and for breakfast the same and then start on his wending way towards the west thinking and hoping to find water in that direction if at all and if per chance he goes two or three days without finding it, he still thinks it is just ahead and pushes ahead as long as hope lasts. (1909)

Charles Mecum:

The old band, who, in 1849 wound their weary way over trackless deserts; tortured by hunger and thirst; slow moving, gaunt skeletons of despair, lost—lost, in a wilderness of desert and mountain never before trod by the feet of white men.

Some gave up the struggle and we buried them where they fell; scooping out the graves with our hands; and covering them with sand as best we could. The rest pressed on, and after much suffering, reached the land we sought. (1901)

If you're going through hell, keep going.

Ed Doty:

[We experienced] Dangers & sufferings & hardships in our wanderings in that dreary desart. i often think of the good feelings that existed amongst the men in all our hardships & trouble, weak & wearry with hunger & thirst, not knowing whose turn it would be next to leave his bones to whiten on the sandy desert. yet thar was no grumbling, no growling, no blaim[in]g was put on any one for leading them into their trouble, but like a band of brothers would do all he could to help his fellow sufferer a long & i do think that theare never was a better hearted set of me[n] eve[r] crossed the plains than the Jay Hawkers of 49. (1890)

Juliet Brier recalled in 1902: "I have been thinking of late more than usual of that perilous journey of 49 and 50 and wonder that insanity did not take possession of us. We endured great suffering of body and more of mind. Four strong men had fallen.... I feared the same fate might befal my husband."

Barrel Springs and William Robinson

Soon they all reached the rift zone created by the San Andres earthquake fault at the northern base of the San Gabriel Mountains, but the incremental movements of these two mighty continental plates as they grind past one another were of no consequence to men who must find water or die. Their lives or ultimate deaths had been peeled down to that single need.

Manly described Barrel Springs nestled among trees in the San Andres Fault when he camped here with the Bennett-Arcan families:

> We reached our camping place in the lower hills at the eastern [northern] slope of a range we must soon cross [San Gabriel Mountains]. Here was some standing water in several large holes, that proved enough for our oxen, and they found some large sage brush and small bushes round about, on which they browsed and among which they found a few bunches of grass. Lying about were some old skulls of cattle which had sometime been killed, or died. (1894, 245)

Barrel Springs was the first watering hole for Indian horse thieves who stole horses and cattle from the Rancho San Francisco on the other side of Soledad Pass. William Robinson died at (or near) Barrel Springs. Accounts vary about how and where he died and was buried. Following are several versions. Edward Bartholomew said:

FIGURE 8.3. Barrel Springs looking east. The Indian horse-thief trail led the Jayhawkers to the springs in the San Andreas rift zone. It saved the lives of all but one—Robinson was buried here. Located on Bear Creek Rd. less than a mile from Pearblossom Highway, Palmdale, CA. Courtesy of Alden Anderson.

I can see in my minds Eye that little party trudging along through that weary desert, so poor and weak that they are scarcely able to throw one foot past the other, but still trudge, finally careless as to himself. We see one sit down by the way side; we leave him to rest a few moments, then follow on. We press on nearly perishing for want of water; finally we reach water; we camp; we rest a few moments. We fill our Canteen, weary as we are. We go back to meet our weak and feeble comrads, to the first we give him a swallow of water & cheer him up by telling him our camp by a spring [is] but little distants. We go on telling [and] giving each as we meet them a swallow of water & telling the good news of our camp by a spring, & finally we reach the last one, poor Roberson too feeble to go further. With help we carry him to camp, lay him down; too late, his spirit has departed, his lifeless form is before us. With our hands, assisted by a stick, we dig his grave beside the still waters. (1884)

James Brier: "Coming to the Coast Range, we found a fine spring on the edge of the desert. Here one of our men named Robinson, lay down for a nap. When we went to wake him he was dead" (1886). "The great sage plain, where we buried poor Robinson, is called Antelope Valley" (1893).

Juliet Brier: "We still remember, as though it were but yesterday, how we toiled on, day after day—hungry, thirsty and in great weariness—sometimes almost in despair, again hoping and believing, though oftimes deceived as to our situation. The loneliness of that night of sorrowing over the death of Mr. Robinson, and also the kindness and sympathy manifested by your company [the Jayhawkers] to our family" (1880).

Manly had still another version after discussing the trip with Shannon and Doty:

One man, a Mr. Robinson, gave out before he could reach camp. He was very weak and had been helped along by putting him on a little mule, holding him on as they rode along, for they were able to help him in no other way. He was pretty near camp when he fell off the mule and said he would not try to ride any farther, but would walk along to camp when he had rested a little. So they left him on his blankets, and as soon as they could get a little soup at camp, they took some back to him, and found him dead. He did not seem to have stirred since he laid down.... As there was not an implement in camp larger than a butcher knife, the body could not be put underground.... [Next day] Sadly leaving this camp and the unburied body of their comrade, they moved up to the edge of the snow [on Soledad Pass]. (February 1889, 190)[19]

Sheldon Young's variation of Robinson's death was mentioned above.

To sum up the various versions—Robinson, on the verge of death or having already died, was brought into the Barrel Springs camp. Some attempt was made to bury his body, probably in the leaves and duff under the trees that drink from the springs.

According to an obituary for Alonzo Clay and a list in the *Beaver Valley Tribune* on February 11, 1898, William Robinson was from Maquon, Illinois, about seventeen miles southeast of Galesburg. (His relationship to the other Jayhawkers is discussed in the introduction.) Robinson's sad demise emphasizes the truly horrific condition suffered by all the northern contingent and how close they were to death.

Over Soledad Pass

The Jayhawker letters do not refer to hiking through snow when they crossed the San Gabriel Mountains over Soledad Pass. However, Manly and Rogers crossed snow on the pass both before and after the Jayhawkers crossed.

Several of the men returned in later years and clearly refer to Soledad Canyon. Dow Stephens wrote Colton, "Do you remember where we buried Robinson? from there we went over a divide and struck on to the head waters of the Santa Clara River [and] followed it down to the San Francisco ranch. the Southern Pacific R R now takes the same route down nearly to the Old Ranch" (1884). Other reminiscences provide additional data to support Soledad Canyon as the crossing site for the Jayhawkers, but the best description is from Lewis Manly who crossed Soledad Pass with the Bennett-Arcan families about a month after the northern contingent:

> The snow belt was five or six miles wide. . . . We must make the best of the cool morning while the snow is hard, and so move on as soon as we can see the way. . . . The glare of the snow [sun] on the sun [snow] makes us nearly blind, but we hurry on to try to cross it before it becomes so soft as to slump under our feet. It is two or three feet in the deepest places, and probably has been three times as deep when freshly fallen, but it is now solid and icy. Our rawhide moccasins protected our feet from cold, and both we and the animals got along fairly well, the oxen breaking through occasionally as the snow softened up. . . . A mile or two down the hill we were at last out of the snow, and a little farther on we came to the little babbling brook. (1894, 245–47)[20]

Thus, the northern contingent finally escaped the Mojave Desert, but their struggle was not yet over.

CHAPTER 9

RANCHO SAN FRANCISCO: "VALLEY OF DELIVERANCE"

For the first time in many days
our little party of breathing corpses broke into a trot.
—John Colton, February 1888

Delivered from the jaws of death and snatched…
from the burning sands of the desert to a paradise on Earth.
—Lorenzo Dow Stephens, 1892

It was like coming back from death into life again.
—Juliet Brier, 1898

Down Soledad Canyon

John Colton wrote T. S. Palmer on April 12, 1894, that "as to the Pass—from what you say, it must be the Soledad pass, as it was 50 miles from the [Barrel] spring, that we followed down the River to the S. F. Ranche." Sheldon Young gives fifty-two miles from Barrel Springs to the rancho *casa* (January 28–February 4, 1850). Mileages figured on Google Maps and National Geographic topographic (TOPO!) maps provide a distance of about thirty-eight miles, but for the '49ers, in their weak and starving condition, following the windings of the landscape and the river, that mileage was magnified. "Well do I remember when we crost the devide and found Grass & water," wrote Urban Davison. "Tom Shanon laid down and kissed the Ground" (1899).[1]

A month later, with the Bennett-Arcan families, Manly observed that the Jayhawkers had camped at the beautiful "little babbling brook," the same place where Manly and Rogers had their supper of crow, hawk, and quail a few days before (1894, 247). "The [Jayhawker] hunters here began to branch off and try to get some better meat.... But when they began to climb the adjoining hills they soon found they were too weak to do much and had often to sit down and rest. When they tried to run they stumbled and fell, and found they were pretty weak hunters" (February 1889, 190). "Plenty sighns of game but none killed yet," wrote Sheldon Young. His diary guides us over Soledad Pass and into Soledad Canyon, but he forgets January has thirty-one days instead of thirty, so his February dates are off by one day. Correct dates are in brackets:

> Jan. 30 went 6m SW [from Barrel Springs] • plenty sighns of wild cattle
> & Deer • Stopt at noon [after crossing Soledad Pass] • cattle took a
> stampeed • Campt no water • Bunch grass Cedar wood[2]
>
> [Jan. 31] Feb 1st left camp at noon • went 8m S.W. • found part of the cat-
> tle • 4 head yet not found • struck into a large Kenyon • a fine steam of
> water • suppose it to be the head of Kern River • it is lined with plenty
> of timber & grass • it looks like home • plenty sighns of game but
> none killed yet
>
> [Feb. 1] 2 went 12m W following Down a Deep Kenyon [Soledad] • it is
> lined with heavy timber • Some places Difficult to get through • there
> is a variety of Species of timber Live oak &c [3]
>
> [Feb. 2] 3 went 13m W • followed an old [trail] what seems to have been
> an old pack trail • found some Difficult thickets to get through • had
> good grass & water • plenty of live oak wood

Sheldon still thinks the northern contingent is on Barney Ward's cutoff and they have just crossed Walker Pass, which is almost a hundred miles north. Six miles southwest of Barrel Springs places the argonauts in the flat area where Kentucky Springs Canyon enters Soledad Canyon.

Just after crossing the snow on Soledad Pass, "the cattle stampeded wildly and broke off into the bushes.... They [Jayhawkers] tried to follow the escaped ones, but the brush was so thick and themselves so weak, that they had to give it up and go on with the one ox and what few blankets he had on his back" (Manly, March 1889, 14:1). Fortunately, they "found part of the cattle," although "4 head yet not found." Years later Tom Shannon was still disgusted with his "bolly ox that skipped with my pack when he found the first green grass in Santa Clara canyon" (January 22, 1897).[4]

In the upper reaches of Soledad Canyon, "Capt. Doty and his companions" shot three wild horses. Finally, the Jayhawkers had meat other than the corruption provided by their emaciated oxen. The horses, "dressed and divided up," were "proving to be excellent eating" (Manly, March 1889, 14). Ed Bartholomew also recalled "the suppers from those three wild horses" (1888). John Groscup wrote, "Wee folowed that trail till some Horses 3 wee found and them wee cild [killed], I never ate Eney thing in my life that tasted as good as thay. this was the first Day of Febuary 1850. thare wee camped one day. the next wee came [where] Dear war[e] plenty and White oak acorns fore Coffa [coffee]" (1885).

Somewhere along the stream, "one place was quite swampy.... Some of Brier's cattle went in for grass and Mrs. Brier could not club them out, and was forced to go after them, and in this expedition she got mired herself and could not get out.... Mr. Stephens came along that way, and seeing her trouble went in and pulled her out of the mud and assisted her to hard ground again. Brier never moved, and neither said a word, but Stephens looked awful cross at him" (Manly, March 1889, 14).

The Reverend James Brier apparently suffered extremely from illness and complications of diabetes. He said that at Barrel Springs, "for six days my wife had had to lift me up every morning, and it was hours before I could command my muscles. She was a delicate little woman, ordinarily weighing 115 pounds, but she stood the horrors of that trip better than any of us.... She had to cinch up the oxen for the men every morning" (1886).

John Wells Brier related what he remembered (or had been told) about entering the rancho lands:

> The grass grew stronger and more varied; evidences of animal life began to appear, and about noon of the second day we killed a mare and two foals. What a banquet they furnished no one can appreciate who has not lived for months on the flesh of diseased and thirst-wasted oxen. Here we also experimented with acorn bread which proved, literally, a bitter disappointment. Our shoes had long ago worn out, and many of us wore moccasins of green raw-hide, which, with our generally ragged outfit and skeleton plight, gave us a very grotesque appearance. (1903, 456)

From the pass to the valley of the Santa Clara River was about twenty-eight miles of rugged, brushy canyon country. Manly: "The hills on the sides were covered with large brush, and on a higher part of the mountain south, were some big trees, and we began to think the country would change for the better pretty soon. We followed down the ravine for many miles, and when this came out into a larger one [where Aliso Canyon enters], we were greatly pleased at the prospect, for down the latter came a beautiful little running brook of clear pure water [Santa Clara River]" (1894, 169).

To protect themselves from the brush, Manly said, "The soles of our moccasins had been worn out for several days, but we had kept them on to protect the top of our feet coming down the brushy canyon" (L. and J. Johnson 1987, 92). The Jayhawkers doubtless did the same.

Many miles of streambed and brushy bank had to be negotiated before the Santa Clara River went underground and only a sandy wash remained. Manly best described the brushy canyon when he and Rogers brought the Bennett-Arcan families through:

> We had the trail of the Jayhawkers to follow, so the vines, brambles and tangles which had perplexed Rogers and myself in our first passage were now somewhat broken down.... [T]he hills came down so close on both sides that there was no room except in the very bed of the stream. There was no other way, so we waded along after the oxen as best we could. Some times the women fell down, for a rawhide moccasin soaked soft in water was not a very comfortable or convenient shoe, however it might be adapted to hot, dry sands.... It was indeed one of the hardest day's work of the whole journey....
>
> The next day we did not wade half as much, and after a few hours of travel we suddenly emerged from the brush into a creek bottom which was much wider, with not a tree to obstruct our way. The soil was sandy and covered more or less with sage brush, and the stream which had been strong and deep enough to make us very wet now sank entirely out of sight in the sandy bottom. The hills were thinly timbered on the left side but quite brushy on the right, and we could see the track of cattle in the sand. (1894, 255–58)

Desert over-and-under rivers and streams, like the Santa Clara, sank out of sight when evaporation and seepage dissipated the aboveground flow in the vicinity of Oak Spring Canyon. This was near the eastern boundary of the Rancho San Francisco, about eight miles east of today's Saugus. As Manly said, "It was astonishing to see how the thirsty sand drank up the quite abundant flow" (1894, 257). James Brier said the stream "disappeared lower down. We came into a beautiful level valley, 3 or 4 miles wide—the Santa Clara Valley [today's Santa Clarita Valley]. There was no more water. I sat down under an oak tree, concluding to die there; but was cheered by finding acorns, which we ground and ate" (1886).

The Jayhawkers and Briers apparently camped before reaching a thumb of land that juts northward into the valley from the San Gabriel Mountains near today's Saugus—Manly's "point of hill that nearly crossed the valley."

A trail crossed this "spur from the hills" to a small spring just west of its summit. The hill overlooks the broad, verdant Santa Clarita Valley—the grazing lands of the Rancho San Francisco. Juliet Brier had to pack up their camp and collect their cattle before she and her boys (plus two other young men) could follow the others:

The morning seemed more dubious to me than those preceding, as it became necessary to help my husband rise from his place of rest and [he] told me he must walk on before as he feared he should not be able to keep up. Trails [from horses and cattle] were in every direction and I feared he might take a wrong one and get lost. He assured me he would keep in sight of the train and started on. We followed as soon as possible assisted by two boys—Patrick and Loomis.... The next time I saw my husband after his early start—when we overtook him—he was standing amid the crowd with a big piece of meat in his hand. His first words to me were "No more starvation, Julie." (1905)

In sight of so many fat cattle, "there was but one object in view at that time, and that was *eat eat eat*" (Stephens 1892). Several Jayhawkers make reference to shooting a cow or bullock. In fact, more than one was shot, and each reference adds a detail that helps complete the picture. First is Sheldon's curt "Wolf... killed one cow got caught at it." Then Urban Davison: "When we got in to the valley Harrison France kild the Heffer with the only pistol in the company. we was eating the raw meat when the Spaniards rode up mounted on Caviyes [Cayuses] armed with Carabines [who] dis Mounted and Jerked the meat out of our Hands" (1899). Edward Bartholomew added: "Neaver a meal that I will remember longer than those we eat just before arriving at Francisco's Ranch" (1882). Ed Doty: "You will see vast fields of waving grain & hear and see the puffing of the iron horse coming in to the little town on or near the same spot whare you shot the wild bullock" (1888). John Groscup: "So on wee went until the 4th and some Cattel then wee kiled and then to the ranch [house] wee came" (1885). Sheldon Young:

> [Feb. 3] 4) lay in camp this Day · two dee[r] killed · one man lost
> [Feb. 4] 5) went 7m W · had Dry Camp · Plenty grass & timber & a wide Kenyon · we are in a country where there is plent[y] of Deer wild cattle & horses · there was such a Bellowing with [?] cattle that there was but [little?] sleeping last night · Wolf [Tauber] went in pursuit of some · killed one cow · got caught at it · found a ranch 6m from Camp · had thousand[s] of cattle sheep horses & we have got out of trouble at last

Young's "one man lost" was a Frenchman who wandered away and was thought to be dead until years later surveyors found him among the Indians—the '49ers did not remember his name.[5]

The camp where the Jayhawkers shot the cattle (Sheldon's "wide cenyon") was six miles from the ranch house, just east of the thumb of land that juts northward at Saugus. LeRoy hiked over this brushy area in 1977 when he and Dick Bush hiked from Death Valley to the site of the rancho casa searching Manly and Rogers's route, but a few years later it was covered with homes.[6]

The report of the 1887 Jayhawker reunion in the *Republican Register* said:

> The boys shot five head of the cattle and were eating the raw flesh and fat when the ranch Indians, hearing the firing, came down with all the shooting irons they could muster, but seeing the helpless condition of the party, they rode back to headquarters and reported to Francisco, the Spaniard, who owned the ranch and cattle. He came down and invited them to camp in a grove near his home; bade them welcome, and furnished the party with meat, milk, grain, and everything they needed, and kept them until they were recruited and able to go on their way. [The Death Valley '49ers did not differentiate between a Spaniard and a Mexican].

The *Kansas City Times* shed additional light on this part of the journey (although with some minor inaccuracies and possible exaggeration):

> Mr. Brier said little of his wife, but all members of the famous party hold her name in reverence. She had with her two boys, aged seven and four years. Throughout all the frightful hardships of the march not once did her courage desert her, not once did she falter. One pony was the only horse in the outfit. This she rode occasionally, but most of the time walked with the men. Nothing could daunt her.
>
> The morning of the day the Francisco ranch was reached not even Mrs. Brier was able to urge seven of the men onward, and they were left behind to meet what seemed certain death. The Spaniards, however, were made to understand that some of the party were behind, and starting out promptly they brought all safely in. Mr. Colton, who is a man a trifle over 6 feet tall and of muscular build weighed but 65 pounds when he reached the ranch.... The march of the Jayhawkers was attended with almost unparalleled hardships and sufferings. (February 5, 1890)

Davison remembered "the Spaniards rode up mounted...and Marched us down to the Ranchera and plased us in a big corrill and put a gard over us and fed us on penolie and Leche" (1899).

The emigrants were not put in a cattle corral. They were provided a vacant Indian summer village site near the rancho headquarters. Father Juan Crespí, a Franciscan priest on the first Portola Expedition, wrote on August 11, 1769, that they "made camp close to a very sizable, big village of very fine, well-behaved tractable heathens, who on our reaching here were encamped within a large pen having only one passage for an entrance" (2001, 375). The expedition soldiers called it "*Rancheria del Corral*, Pen Village," with "at least five hundred souls here, what with men, women, and children." Since the Jayhawkers arrived at the rancho in winter, the empty summer habitation site, with its brushy enclosure, offered protection for the weak and malnourished emigrants from wandering cattle and bears. The casa could not hold them; thus, the Rancheria del Corral was the best protection the Mexicans could provide.

Frances Mecum, wife of Jayhawker Charles Mecum, was not a '49er, but she wrote the following account in chapter and verse similar to a biblical story based on what Charles told her:

> [Chapter I, verse 8]...They one day entered a pass in the mountains:—and suddenly emerged into a beautiful valley:—on which fat cattle were grazing:—and the grass, and wild oats, were abundant.—
> Chapter II
> 1 And they wept for joy:—for they thought they had found the lost Eden:—and no angel, with flaming sword was there to prevent them from entering, and partaking of the abundance before them. And they killed fat cattle, and ate of the raw flesh.
> 2 And the hirelings came down with ancient weapons, and fain would have driven them away:—but they budged not. And they returned to their master, and reported. And he was a noble Samariten:—Francisco by name:—and he came among them, for *his* was the Eden, and the cattle on a thousand hills. And he rebuked them not.
> 3 And when he saw their starved, and wretched condition, he welcomed them to his ranch:—and gave them milk, and grain, and meat:—and they sojourned with him till they were strengthened and able to go on their way. And they rejoiced greatly at their deliverance:—and dwelt again among the children of men. (1889)

The Indian vaqueros reported to their Mexican boss who came to see these creatures who where killing his cattle and invading the property under his care. José Francisco de Gracia López, age forty-seven, was indeed the principal resident, manager, and foreman of the Rancho San Francisco, and his livestock grazed on its lands side by side with the cattle and horses of the del Valle owners. Francisco was a maternal uncle of the principal landowner, Jacoba Feliz del Valle Salazar. Francisco and his wife, Antonia, had lived in the casa for many years. He was the rancho's man on the land, while Ignacio del Valle was the man in the office in Los Angeles. Ignacio was the literate oldest son of Antonio de Valle and inheritor of part of the rancho. He was several years older than his stepmother, Jacoba (J. Johnson 2018).[7]

Santa Clarita Valley Proper—The Rancho San Francisco

The eastern boundary of the rancho was roughly twenty trail miles west of Soledad Pass about one and three-quarter miles up-canyon (east) from where Sand Canyon enters the Santa Clara River wash (G. H. Thompson 1874). This boundary is about eleven air miles east of the rancho casa, which is on a bluff overlooking Castaic Junction.

The surviving '49ers from the northern contingent—Jayhawkers, Brier family, Young's mess, a couple of Germans, and other assorted men—entered the heart of the Rancho San Francisco on February 4, exactly three months after they turned west from the Old Spanish Trail into unknown territory to follow Barney Ward's bogus, supposedly well-watered, shortcut to the goldfields.

James Brier recalled the event:

> Soon one of the boys discovered a ranch up on a hill, where the Santa Clara makes a turn to the south, and near where a tributary stream [Castaic Creek] runs in from the north. An old Spaniard came out with his Indian servants armed; but, on seeing who we were, and how nearly dead, welcomed us with great kindness. When my wife and boys came up he broke down and cried.
>
> When we came to the ranchhouse, his wife and daughters fell upon my wife's neck and nearly smothered her with tears and kisses.[8] The old Spaniard killed a bullock and put it where we could help ourselves. We had no money, but he said he wanted none. Starved as we were, we ate heartily and came near killing ourselves. My little four-year-old boy who had tramped 400 miles, bloated up fearfully and would have died but for Dr. Irving, who happened to be there from Los Angeles. (1886)

Dow Stephens, writing in three different letters, said:

> [1886] When we reached the San Fruncisco Ranch, that long to be remembered Oasis, was there ever such a sight! was there ever such Joy among a band of walking Skeletons before, just emerged from out the great American desert and Death Valley!
>
> [1892] I can look back now to that day we entered that beautiful little valley, the San Francisco Rancho clothed in its richest verdure and wild flowers with here and there a Stately Oak and dotted with innumerable herds of Cattle, which was the principal factor of interest to a Starving Jayhawker.

[1896]. [This was] the day we were delivered from the jaws of death and snatched, as it were, from the burning sands of the desert to a paradise on Earth where the fatted calf was Slain and we did eat thereof and was filled. Well do I remember the Joy that was Manifest in that Camp that night for we knew we were safe and had Struck a land of plenty.

Manly recalled:

A most pleasing sight filled our sick hearts with a most indescribable joy.... There before us was a beautiful meadow of a thousand acres, green as a thick carpet of grass could make it, and shaded with oaks, wide branching and symmetrical, equal to those of an old English park, while all over the low mountains that bordered it on the south and over the broad acres of luxuriant grass was a herd of cattle numbering many hundreds if not thousands. (1894, 272–73)

Charles Mecum:

After wandering *so long*, on the trackless desert—where all nature seemed dead and lifeless—*no* grass,—*no* trees—*nothing* to indicate life;—when, *just 28 years ago today*—we entered that beautiful valley—*luxurient* with grasses, & verdure, —the air redolent with the fragrance of flowers—thousands of fat cattle grazing—untold abundance for man, and

FIGURE 9.1. Depiction of Rancho San Francisco casa looking southwest. Santa Clara River is to the lower right. Arthur Perkins labeled it "Asistencia of Mission San Fernando, 1804. Artists reconstruction by Adolph Henkel, 1957." ABP-3-14, courtesy of Arthur B. Perkins estate.

beast—how our hearts overcame us. All our sufferings had not brought a tear—yet *then* we wept like children for joy.... Shall we ever forget the good Samaritan, who cared for us till our strength returned? We were *strangers*—yet he took us in:—*hungry* and he gave us food:—*thirsty* and he gave us drink. We were *poor*:—we could not repay him:—but in the summing up of all things he will receive his reward. (1878)

Mecum to the 1901 reunion:

We long before had lost all thoughts of gold:—food, water, and life was all we sought. And there in that glorious land, we found life, food and rest:—surely Heaven can seem no more beautiful to world weary mortals, than did that beautiful valley of Santa Clara which was an Eden of flowers, verdure with many streams of pure water, and thousands of cattle grazing there.

Some said, "Here we found life and deliverance, and here will we dwell." Your party of survivers to-day, will be made up of those who wisely remained in that land.

Juliet Brier remembered "the bright and glorious morning that ended those months of doubt and starvation, and opened to our vision once more green fields, cooling streams and plenty to satisfy the cravings of our starving, famishing bodies; and for the gold hunters, the brightest hopes. There our companionship ended, and we were scattered."[9] Years later she recalled, "We in our joy sorrowed for those who had fallen in the desert and those scenes are fresh in my mind to day" (1906). Several Jayhawkers did remain in this golden land, settling northward from Santa Barbara.

Sheldon Young summed up their deliverance succinctly: "We have got out of trouble at last."

The Rancho San Francisco is an example of California history in miniature from its native American villages, through Spanish exploration, the mission period, Mexican land grants, the first California Gold Rush, cattle ranching, oil exploration, and, today, urban sprawl. A history of the rancho is a study worthy of its own book.

The adobe *casa* was built on a bluff on the south side of the Santa Clara River west of the mouth of Castaic Creek, a rectangular building surrounding an interior courtyard. The roof was made of reddish tan clay tiles shaped over the thigh of the tile maker, and the floor was composed of two-inch-thick flat tiles of the same local clay baked hard in a wood-fired kiln.[10] There was a corral nearby where the Indians could tie a wild cow to procure milk for the needs of the

FIGURE 9.2. Rancho adobe milk house (cooler for milk, butter, meat) in the 1930s. Off the picture left was the casa, and the Santa Clara River is down the hill. ABP-13-22, courtesy of Arthur B. Perkins estate.

household. To keep the milk, butter, cheese, and fresh meat cool, an adobe milk house was built over a spring in the shade of some cottonwoods and California live oaks just over the bluff.

Several of the '49ers expressed their grateful feelings toward the people at the rancho. Urban Davison: "our little Party was cast in a foran [foreign] clime and thare surrounded with People who spoke with different tongues. how kindly we ware treated" (1883). "This was San Francisco ranch and there we were kindly provided with food in abundance," Juliet Brier remembered in 1905.

When Manly and Rogers were preparing to return to rescue the two families in Death Valley, "the woman came out with four oranges in her hand and pointed to her child and by signs gave us to understand we must take the fruit to the children and not eat it ourselves" (L. and J. Johnson 1987, 98).[11] For many years researchers (including LeRoy and me) have assumed the lady who brought the oranges was Jacoba Feliz del Valle Salazar, inheritor of the largest share of the Rancho San Francisco—the woman who lived on the rancho with her husband, Antonio del Valle, from 1839 until his death in 1841 (J. and L. Johnson 2013).[12] But now we know Maria Antonia López, wife of Francisco López, was the woman who was "disposed to be very friendly." Antonia fed Manly and Rogers and gave them the oranges. She had lived on the ranch with her husband since 1835 and was still there in 1850 when she opened her home to the starving emigrant families.

Manly recalled, "Aside from her dark complexion her features reminded me of my mother" (1894, 184). Antonia López, age forty-eight in 1850, was about the age of Manly's mother, whereas Jacoba was the same age as Manly—both born in 1820. Juliet Brier (1898) said, "The old lady burst out crying when she saw our condition," and John Rogers said, "The old Spanish lady took the two families [Bennett and Arcan] to the house and kept them there two days" (1894; J. and L. Johnson with Faye and Ryan 1999). Neither "old lady" description fits Jacoba, who was six years *younger* than Juliet and only two years older than John Rogers (Newmark and Newmark 1929, 59; Manly in L. and J. Johnson 1987, 154).

Manly also spoke of the hospitality extended to the Bennett and Arcan families when they arrived at the rancho more than a month after the northern contingent:

> [Our] hosts first baked some kind of slapjacks and divided them among their guests; then gave them beans seasoned hot with pepper: also great pieces of squash cooked before the fire, which they said was delicious and sweet—more than good.... The children were carefully waited on and given special attention to by these good people, and it was nearly ten o'clock before the feast was over: then the household had evening worship by meeting in silence, except a few set words repeated by some in turn, the ceremony lasting half an hour or more. Then they came and wished them *buenos noches* in the most polite manner and left them to arrange their blankets on the floor and go to sleep.... When we made a sign of wishing to pay them for their great kindness they shook their

FIGURE 9.3. The location of the casa milk house was on a narrow bench below a larger bench where the casa was located. It overlooks the Santa Clara River. 2010. Leaning tree might be the small tree in the Perkins photo.

heads and utterly refused. It was genuine sympathy and hospitality on their part, and none of us ever forgot it. (1894, 263–64)

John Brier also referred to the rancho folks:

We were at the base of the hill on which stood the adobe house and cor-rals of the Rancho de San Francisquito.[13] Presently two vaqueros led in a coal-black bullock and slaughtered it for our use. Others rode down from the house with squashes, beans, corn-meal and milk, and there were tor-tillas for immediate eating. We fed like hungry animals, and some of us would have died from the sickness which followed, had not Dr. Irving, of Los Angeles, arrived most opportunely. (1903, 459)

The effects of suddenly having sufficient food to eat played havoc with the diges-tive systems of the '49ers now adapted to near-starvation conditions. They could not adjust to the fat and vegetable roughage and rebelled at the onslaught with bloating, vomiting, and considerable internal pain as their bodies tried unsuc-cessfully to absorb it.

Urban Davison recalled in 1878: "It has been 27 long years since our small party landed on the pacific coast & was so kindly treated by the Spaniards, when John Colton & my self partook so sumptously of the spanish suit [suet] & groned all night to pay for the same." In 1899 he wrote, "John Colton & My self went with the Spanyard to this wickiup. we helpt ourselves to suit [suet] that was Hung on a line and fil[l]ed our pockets un be known to the Spaniard that was Guarding us. we did not mean to steal it. we jest took [it] to stay our craveing apatite. we paid dearly for it in a bout two Hours after it. we was taken very sick. Oh how I did Hate to own up to taking the suit." Women at the *casa* would have asked to have the best suet (the hard fat around loins and kidneys of beef or mutton) saved separately when a bullock was slaughtered because its high smoking point made it valuable for frying and baking. Thus the guilt Davison felt for eating it.

Edward Bartholomew suffered the same way: "Those eight days we lay there suffering. Ah: I can hear those groans [s]til yet from all parts of that Camp & more particularly from that mess under the tree where Our Captain [Asa Haynes] lay" (1885). The Bennett-Arcan families experienced the same symptoms from their craving for fat. They had to camp suddenly when Sarah Bennett was "taken sick with severe pain and vomiting.... The rich, fat meat was too strong for her weak stomach" (Manly 1894, 252–53).

Edward Bartholomew summed up the Jayhawkers' harrowing trek through the Nevada and California deserts when he said: "Memorable 4th of February 1850, the day that little band of lost *Gold Hunters* past, as we might well term it, from death unto life or from a state of starvation into a state that flowed with milk & honey. The day of our relief from starvation, the day of our deliverence, the day we should celebrate" (1872).

Journey to the Gold Country Resumes

When the Jayhawkers recovered sufficiently to move on to the goldfields, most of them headed down the Santa Clara River to the coast. The Brier family, Goller and his friend, and possibly others headed south for Los Angeles or the port of San Pedro: "We camped at the ranch two or three days," said James Brier, "and then we came over the San Fernando mountains to Los Angeles. I came in here with a pair of old bootlegs tied on my feet with rawhide. I weighed 75 pounds. It was Saturday evening, February 12, 1850. Here we found those who went with Hunt. They had arrived in December after considerable suffering" (1886).[14]

Sheldon Young's mess took a ship to San Francisco from either San Pedro or Santa Barbara. At Santa Barbara, some of the Jayhawkers took "a boat that was there loading hides, while the group that Doty stayed with traded their oxen for horses and continued north into San Luis Obispo (County) and to Stockton from where they went to the mines" (H. Doty 1938).

John Colton wrote about leaving the rancho:

> I was one of the lost party, the "Jayhawkers of '49," that got out to the desert and found the first food at the Rancho San Francisco, on the Rio Santa Clara, about twenty miles east of the mission of Santa Buena Ventura, on February 4, 1850. After two weeks of appeasing starvation appetites, we began to recover and consider how we would get to the mines six hundred miles away. A party of Mexicans down from the mines were about returning and wanted me to go with them. They gave me a good horse, saddle and trappings and we left the ranch about February 20, the balance of our party got to the mines by various routes.[15]

After most of the Jayhawkers headed west down the Santa Clara River and neared the Pacific Ocean, they entered the Mission Buenaventura lands. Ed Doty mentions an incident with the mission alcalde: "The iron horse comes rite on down the [Santa Clara] valey to Ventura, whare the alcalda stop[p]ed us for killing a fat cow a short distance from that place, which we setled by paying 5 dolla[r]s, & from ventura the old puffer comes [north] to Beautiful santa Barbary, the charmed city by the sea" (1888).

CHAPTER 10

AN INCOMPLETE AFTERWORD

The Men who moil for gold.

—Robert Service, "Cremation of Sam McGee"

A SEPARATE BOOK could be written about the Death Valley Jayhawkers after they recuperated at the Rancho San Francisco. All ultimately went to the goldfields (although the Reverend James Brier went to minister to souls, not to dig for lucre). Some stayed less than a year in the golden state (for instance, Haynes, Tauber, Young), and some stayed more than two years before returning east to their family homes or to settle near where they had been brought up (Clarke, Clay, Colton, Gritzner, Haynes, McGowan, Mecum, Morse, Richards). Some died in the gold camps (Cole, Larkin, Palmer). Some ventured into the desert again looking for the elusive silver mountain (Brier, Rude, West). A few were seriously struck by gold fever and became part-time, or even lifelong, miners from Arizona to Alaska (Davison, Frans, McGrew, Rude, Stephens, West). Several served in the Civil War (Bartholomew, Byram, Colton, Shannon). Others settled in California in the established patterns of farming or commerce (Allen, Doty, Groscup, Shannon, Stephens, Woolsey). Names given are illustrative, not all-inclusive. My intent here is to provide an overview, with a few colorful details, of the Jayhawkers' activities in the gold diggings and the variety of their lives after surviving their extraordinary struggle through the western deserts that included Death Valley. Much information for this chapter came from genealogy, military, and Census websites.

Some of the Jayhawkers had commented on trying to forget their harrowing experiences in the deserts, but as time went by and reunion reports circulated, they began to share their experiences and find solace by putting reminiscences on paper.

The *Galesburg (IL) Republican Register* summed up in 1895, "Some of them have an abundance of earthly store while others, though growing old have still to labor hard for a livelihood."

Some of them wrote home from the gold camps. Years later John Colton wrote about his travel up the coast after the desert trek. At age sixty-six, he recalled being the savior of the day when he was an emaciated gold-seeking lad of eighteen while on his way to the goldfields with some Mexicans. It gives a snapshot of early California:

The weather and scenery along the trail up the coast was magnificent and thousands of fat cattle always in sight. The old California law allowed persons to take what beef was necessary to appease hunger. My Mexicans had good horses and lassoes, and were strictly law-abiding people, and we lived on the fat of the land. In the first part of March we reached San Jose. The first legislature of California was then in session and had just elected General Fremont United States senator.[1] The rain commenced falling soon after our arrival and continued for nearly a month so we could not stir a hoof and were forced to abide the time when it would stop, much against our will, as we missed the unlimited free beef of the trail, for with thin purse and provisions at a terrible price, we had to put up with our adobe hut, and wait for better times. The first legislature, as you may know, was a hilarious body, whose daily duty seemed to consist of meeting in the forenoon and adjourning to "La Mina de Oro" to take a drink. "La Mina de Oro" was an adobe building, the grand roundup for the first-class citizens of the capital to drink, gamble and talk politics.

One morning I was at the roundup and heard some parties talking of a coming election, mentioning the name of John Yontz for sheriff; the name sounded familiar to me, and after a moment's reflection, remembered that he formerly kept a tavern in Peoria, Ill., where my father had teams put up that hauled barrelled pork from his packing house in Galesburg for shipment to New York, via New Orleans and the ocean. I soon found him and he remembered me, although I was but a boy then, and when he told me that he was hard pushed for votes for sheriff and the office was a rich prize, I told him I could help him out. Every thing that could ride a horse could vote, regardless of his place of residence. There were a great many camps of Mexicans all around the outskirts, waiting for rain to quit, and being with Mexicans I was well known to all of them, and "El Muchacho" was always welcome in their camps. A little explanation to them that my friend whom I wished to be elected needed all their votes, together with a liberal distribution of aguardienti, resulted in a tremendous vote from my troops and Yontz was triumphantly elected. The celebration of the victory following was of grand proportions, assuming later in the night a lurid hue, transcending anything that had transpired in that new State. The rain stopped soon after and we were in the saddle and off to the placers. (January 30, 1897)[2]

Dow Stephens (1916, 28) wrote about the cost of mining in March 1850: "My first experience in mining was at Merced River, where I paid sixteen dollars for a shovel, eight dollars for a pick, fifty dollars for a rocker, four for a gold pan, and thirty-two dollars for a pair of boots. Everything else was in proportion, and vegetables were out of the question, as I saw a man pay a dollar and a half for a single onion." Lewis Manly also wrote about the mining camps:

In '50, Georgetown, situated on the divide between the American river and Canyon creek, some of the business houses used by gamblers were big tents occupied by that gentry and their lady friends, who were in some respects scantily clothed, and they were gazed upon by many a curious miner, who had been out of female society for a long time. . . .

In '50 this [Illinois Canyon] was a lively mining camp; the claims were 15 feet square and 10 feet deep, and every claim was stripped to the bedrock and some water found, but not hard to contend with. The gold was close to the slate bedrock, and the best of it was very bright and about the size of water and musk melon seeds. The boys camped here were mostly good jovial fellows, making pillows at night of their books [boots?] and living mostly on slapjacks and beans, and the first one up in the morning at the peep of daylight would give a loud Indian warhoop, and would be immediately answered from other camps from one end of the canyon to the other, and their shrill voices would echo from up and down the canyon, where all was still silent. . . .

Then the smoke from the many camps would ascend straight up in the still air until it vanished in the tops of the lofty pines, and soon the sound of the pick and shovel would ring in the clear, pure balsam-laden air, that gave health and vigor to all busy pioneer boys, who had just walked from the East to the Pacific Coast, and all were eager to work and fill the leather purse they brought so far. (1896)

On July 28, 1851, Colton wrote his father from the rich diggings around Sonora:

I thot I would write a line to day as I was in town & had a chance. I wrote you a long letter a little while ago telling you that I was just going into

FIGURE 10.1. Gold washers of California. Hand-tinted lithograph, 1853. Johnson collection.

a Co[mpany] to turn the River Stanislaus into the dry diggings. Wall [Well]—we have worked 18 days (162 men) & have made a grade six ft wide—6½ miles [long]. 2½ more will finish which we will do this week. We found it impossible to dig a ditch that would hold water on ac [account] of the looseness of the soil and the mts were straight up and down. The point where we commenced was 2000 ft above the level of the river. Now we are coming down into it fast. After finishing the grading we have to floom [flume] it. We are going to begin the saw mill tomorrow with 3 saws besides 25 whip saws to saw by hand. There is plenty of good pine timber. We shall make the floom 4 ft wide & 2 ft deep with 12 ft fall to the mile which will rush the water in mighty fast—When we get the water to the summit we divide it and send it in all directions. It is mighty hard work. The soil is so light that there is a perfect cloud of dust all the time there. The hot sun shining right on the side hill is enough to kill a Missisipi nigger to say nothing of a white man. Then we have to pack our water a long distance and half the time choked to death. Shares already sell for from 2[00] to 400 dollars. This will be the best property that I know of. The Co all have their minds made up for 5000 Each per Anum for the use of water but I do not fly as high as that. But I am ceartain of one thing, that I can sell out & with what I make in the mines have 5000 dollars the 1st of next June [1852]. Wheather I shall sell or not I do not know. It will bring in a mighty [blank] Salary to a person living in the States and

it will be a thing that will last a good many years. The least that will be
charged for the use of water will be 5 or 6 dolls a Tom pr day and we can
furnish enough to run 10,000 Toms besides Saw & Quartz mills, for the
use of gardens &c &c.[3] But then if a person is not here to attend to & recv.
his money, the agent will Swindle him, if they are not men that would
not steal if there was no punishment—Business is brisk in Sonora now,
& when the water gets through business will be better than the first day
they was discovered—Quartz is turning out better & better. Some Co[s].
taking out as high as 15000 dolls pr day.

I have not rec'd a word from you since Feb last. I wish you would write
often and let me know how things are getting along at home. The boys as
far as I know are all well. Gambling & Lynching go on as usual—Give my
respect to all hands.... After this, direct your letters to Sonora and I think
I shall get them. Send me a paper once in a while.

A month later Colton again wrote his father, this time from San Francisco:

I am in good health & have as good a sight for making my pile as ever....
I am here now to help run a small steamboat up to Stockton [closest port
to the Sonora gold country] for the purpose of getting the engine for a saw
mill to make our floom. I have spent some money & some time & expect
to have to spend a great deal more. Some have quit, sold out for little or
nothing & left on a/c [account] of its taking so much more time than they
expected. But I am going to stick to it untill the last trump is played, for I
am ceartain of one thing, that it cannot fail to pay me the common wages
of labor 100 or 125 per mo. for all the time I spend clear, with the chance
of considerable more. And if it does finaly prove a failure I can work a mo.
[month] or so, get money & pay my debts & then have as good luck as I
ever have had. But we do not calculate to finish the whole distance this
year but bring the water from the South fork which as soon as the snows
& rains come in the mts, will furnish sufficient water to fill a floom 4x2....
If I do well in the mines this winter and I can sell [my flume share] in the
spring or sumer I shall come home, if not I shall not.

The boys are all scattered about the mines. I dont think they are doing
much. If I could only have had water last winter & this summe[r] I should
have made my pile but so it was'nt. Old Sid [Sidney P. Edgerton] is somewhere in the northern mines. The last time I heard from him he was trying
to pay his grub. He has borrowed some 150 or 200 dolls I suppose of the
boys. If I were in his place I would go home, and send out his young one[?]

for he has got more sense & judgement than he has. I rec'd a letter from you & mother yesterday—July—I have rec'd but one before since last Feb. I hope your health is better than it was then. You should not do so much business.

At present there is considerable immigration. The green ones are awful sick both at San Francisco and the mines, and instead of making their piles they have to hire out for 100 dolls per mo. which does not count up very fast when they are spending anything. If the water could only be in now Everybody would do well. I have claims that would pay 1 oz per day regular.... Write soon.

Yours affectionately, J. B. Colton. (September 11, 1851)

Information about the early whereabouts of several Jayhawkers comes from Charles Mecum at Bidwell's Bar, Feather River, 170 miles north of Colton, in March 1851:

A[lonzo] C[ardell] Clay when he left here was enjoying good health and in good spirits as you ever saw him on Cedarfork. he left here Jan the 3 to go to Red[d]ings diggings. he intends to remain there till it was practicable crossing over via to the heads of Scotts & the Clamoth [Klamath] rivers [in northwest California]. he intends staying up there this Summer and trying his luck at Mining there. I intend to work on the feather [River] or some of its tributaries till next fall. then A C Clay meets me here again and our next move will be our first consideration. I see some of the boys that came through with us. they told me that the deacon R[ichards] had bin verry sick but that he had got well and gone to San Hosa [Jose] and was doing well. Harry Frans is there with him. Urbin Davison is here with me.

In 1854 Colton sent a letter to **Robert B. Taylor** in Stockton and asked the whereabouts of Bill Rude, to whom Colton had loaned some money. Taylor responded:

My Dear John,
It affords me pleasure to be able to acknowledge the ret. [receipt] of yours of the 20th June—I am thereby made acquainted of your safe arrival in Galesburgh and a reunion with your *parent* and my old Friend. You may recollect such was my advice to you when you was toiling at the Big Mariposa in '52. I am most happy to congratulate you in your Mercantile Enterprise.... In regard to **Bill Rude**—I am sorry to inform you that after you left he became Engag'd in Horse Racing and lost not only his money but his Caracter for Honesty. Since which he has not shown his face in Stockton.

A few years later, there was correspondence between Colton and Rude—Colton

FIGURE 10.2. Possible self-portrait showing how skeletal looking William Lorton was after his cross-country ordeal. His rear end is skinny, his mouth caved in, his ribs showing, his legs thin, but he has plenty of hair on his head. "Sierra Nevada mountains/Feather River/ April 30th 1850." BANC MSS C–F 190 v. 8, courtesy of the Bancroft Library.

asking for repayment and Rude putting him off: "Johney we have a fine grasing [grazing] country here & a pleasan[t] one But Bad Society. I am ra[i]sing Stock & have some very fine orchards & vinyards.... I have Don[e] well since I came here [to Tucson] & have often thought I would send you a Draft, But when I had the money I could not get a draft, then when I could get a draft I had no Ready cash" (1859; see also Weight 1959, [32]). By 1861 Rude had "ben Ranching for (4) four years." He sold his first ranch and his interest in the Sopria mine for about $6,000. He wrote Colton, "I have always calculated to sattisfy you for what I owe you.... if you was paid I would not owe one cent in this world & I have property here worth twenty five thousand *Dolls* which I have made By my own labour in the last 4 years & a half.... Now I have no excuse, I have it" (n.d.; see also Weight 1959, [33]). In a later letter, Rude tells us how much that debt was: "I sent you in 1861 April 30th a Government Draft for three hundred Dollars & yet I have not had a Recpt of it alth[ough] I have writen you Several times" (1867; see also Weight 1959, [34]).

Dow Stephens saw William Rude in Arizona about 1863 where Rude showed him "gold nuggets which ranged in size from a grain of corn to the size of a person's thumb. I am quite sure he told me the truth but I came back to California and he got drowned in the Colorado River so the gold is there yet for some future prospector" (1896).

In April 1869 Rude returned to the Panamint Range in company with three men to prospect for the famed Gunsight silver lode and to find the gold coins some of the '49ers buried in a blanket in the upper reaches of the Panamint Range (Miller 1919, 56–64). They left empty-handed (L. Johnson 2013, 289). Rude's sister, Mrs. C. B. Ashby, wrote Colton in 1895, saying that Bill "was drown in the Colorado River in Arizonia Teritory on April 29th 1871. he left a very nice estate worth 50 or 60 thousand dollars" (Ashby 1895; see also L. Johnson 2013).

Some of the Jayhawkers were still keeping in touch from the goldfields in 1854. Tom Shannon wrote Colton: "I received your letter, but not until I had discov[e]red that your mule was at quintaz [Quincy?] and had wrote you to that effect, at San francisco. I have however complied in part with your request. I have sold the mule for ninety dollars which was the best I could do.... if you are not in want of the money I would like to use it for three months at interest" (1854).

John Burton Colton returned to Galesburg after about four years in the gold-fields and worked with his father, Chauncey, and brother, Frank, in the Colton Mercantile store until 1866 (except for his short stint in the Union army when he became very ill and was ordered home). John held the rank of captain and served as quartermaster of the Eighty-Third Illinois Infantry and later as assistant quartermaster of the Union army (Heitman 1903, 318). He had his own store by 1867 at an excellent location at the corner of Main Street and Public Square (Jensen 2007, 22). John married Elizabeth McClure in 1857, but sadly she and their second child died of complications from the child's birth, and their first son, Chauncey, died about three years later. In 1868 John married Mary A. Thomas, and they had three sons.[4]

A picture (daguerrotype), purportedly of John Colton, taken "at Long Wharf, San Francisco, in 1850" after he emerged from the desert, was being sold in San Francisco as "a girl miner in boy's clothing" for five dollars each (Stephens 1916, opp. p. 64).[5] A newspaper article declared after the 1898 reunion:

> Among the pictures of the Jayhawkers is an old daguerreotype of Mr. Colton.... The picture is now dim and faded.... Mr. Colton was but a lad of seventeen [eighteen] at that time, and his pinched and woeful expression gave him something of a feminine appearance. A number of months after his picture was taken he was again in San Francisco and found that the photographer was selling copies of the picture as that of a girl in disguise, and was reaping a harvest at $10 per picture. (*Beaver Valley Tribune*, February 11, 1898)

Note the inflated price! Colton added a comment to the folder containing this heavily scratched picture of a young person with a slouch hat: "Typical miner of '49 San Francisco Oct. 1850." I assume another photo of Colton wearing a vest was also taken in California to send to his family (see portraits).

In 1884 Colton bought a couple thousand acres of ranch land about forty miles northwest of Kearney, Nebraska (near Eddyville), where he successfully raised cattle and hosted the 1906 and 1910 Jayhawker reunions. About 1890 John owned the Kansas City Loan Company, but he returned to Galesburg by 1912, where he hosted four of the next seven reunions (see Appendix C). His portrait is annotated 1893, twenty-six years before he died on October 23, 1919, at age eighty-eight. He died in the hospital at Grand Island, Nebraska, from "injuries sustained" "following a railway accident" while traveling between his cattle ranch, Buzzard Roost, and his home in Galesburg (*Daily Illinois State Register*, October 27, 1919; *Aberdeen-Angus Journal*, Jan. 10, 1820; Nystrom 1936, HM50808). He is buried in Hope Cemetery in Galesburg.

Charles Mecum married Frances Richards, sister of Jayhawker Luther Richards, after returning from California in 1854, where "he acquired a snug lump of color." The following information comes from an account written by Charles's son, Edwin W. Mecum, in 1931. Charles and Frances lived for the next twenty-one years on a successful seventy-acre farm in Woodhull about eighteen miles north of Galesburg. There they reared five children. Charles tried to serve in the Union army in the Civil War, but he had recently broken his leg and was not accepted. To provide more acreage for his growing boys, Mecum bought a two-hundred-acre farm about three miles from Rippey, Iowa. Edwin Mecum wrote this laudatory synopsis of his parents:

> [Charles] was six feet tall, he weighed 185 pounds and was quick and active. Few could cut down a tree as quickly as he, and he could make it fall where he wished.... My father had large, keen, but kindly eyes.... He had a vivid memory in early and middle life, and could repeat poems and pages of prose which he had learned when a boy.... His judgments were charitable, considerate, and generally wise. He was at ease in conversation, and he sought the association of the best men in the community....
> [He] was stern, just, square, and rather quiet. Mother was affectionate, talkative, lively, sympathetic and cheerful. They worked together and celebrated their Golden Wedding Anniversary—Fifty Years—over a year before Father passed away [on February 20, 1905].

In 1905 the Mecums hosted the Jayhawker reunion in Mount Vernon, Iowa, and two weeks later, the day after Charles's death, Frances Mecum wrote Colton, "He had kidney and other troubles which were *incurable*. He did not want to live on account of those troubles" (1905).

Luther Abijah (Deacon) Richards was thirty-one when he went to the California goldfields. He return to Galesburg in 1859 and bought farming land in Woodhull next to his good friend Charles Mecum. Luther married and over time fathered five children. In 1882 he moved his family to Nebraska, and in 1893 he moved to a farm west of Beaver, Nebraska, close to his son George's farm. He hosted six of the Jayhawker reunions from 1875 to 1898 and died on June 15, 1899 (obituary; findagrave.com).

Bruin Byrum also returned from California. His photo was taken in 1864, and Colton's note on it says he died April 11, 1865, in Keokuk, Iowa.

Charles Clarke apparently returned to Henderson, Illinois, where he died September 9, 1865, according to his photo, a copy of an old daguerreotype in which he is holding a pistol.

Alonzo C. Clay returned to Galesburg after about three years in the goldfields and became a successful farmer, family man, and very active leader in the Galesburg community. He served on the town's board of supervisors for fifteen years, was a director of the Farmers & Mechanics' Bank, and was first president of the Knox County Farmers Institute. He also hosted six reunions from 1873 to 1895. He died at home December 13, 1897, at age sixty-eight "after a prolonged illness" (obituary, December 14, 1897).

Edward Doty wrote from San Jose, California, in 1872:

> I do not know of but one of our old crowd in this country. that is Thomas Shanon. I see him occasionaly in San Jose. Crackey is ded & I heard that Dow Montgumery was ded. **Larkin** or Priest as we call him, the Last that i heard from him was living in the north part of the s[t]ate [died in Humbolt County in 1853 (Wheat 1939b)]. George Allen i have not seen nor heard from for several years & Levert West i have not heard from him & know nothing a[s] to their whareabouts. as for my self i have remained the most [of] the time in Santaclara Co. i have me a little home 1½ miles from San Jose whare i am trying to make a living.

Three years later Ed Doty wrote from San Jose: "**Dow Montgomery** was here last October." In 1890 Lorenzo Dow Montgomery wrote a letter to which Colton added two notes: "L D Montgomery with us to Little Salt Lake [actually Mount Misery], but went back to Spanish Trail & to Los Angeles," and he was "Special Agt Indian Depredation Claims."

Tom Shannon wrote in 1879 about seeing several other '49ers: "L. D. Stevens lives in Santa Clara. he was here to visit me a week ago. Ed Doty lives at Santa

Barbary. I have not seen him for a year or more. **Ed Mcgowan** is dead (See McGowan 1872). Norman Taylor I believe is in San francisco. Louis [Lewis] **Manly** lives here at San Jose. you will remember him who in company with the Giant Rogers came out in advance and took provisions back to the families of Wade Arkane and Bennett."[6] McGowan was a Platte Valley Jayhawker who took the Old Spanish Trail.

In 1894 Manly had news about **Lew West**: "I had a letter from him when the book [*Death Valley in '49*] was nearly finished & he said he wanted a copy when he could spare the $2.00 to pay for it. So I conclude he is not rich after shoveling gravel for 45 years." Two years later Manly sent the Jayhawkers the following: "John L West,...long supposed to have passed over the divide, has been found to be living at Phillipsburg, Montana, from whence he sent Colton his photo in 1895. He is now seventy-eight years of age, and writes that his 'hand is shaky' and his health not the best, but could now go and locate the lost 'Gunsight lead in Deth Valley'" (Manly, May 1896).[7]

Colton then searched for West and learned "from Dr. White of the [Sacramento] County Hospital that West, the missing 'Jayhawker,' was a patient there recently, but that he died on the 12th of last January" (*Sacramento Record-Union*, April 1898) about eighty-three years of age. Over the years West had been reported as mining in British Columbia, Idaho, and Coloma, California.

The following information about **Edward Doty** comes from an interview with Ed's son, Henry Doty, age eighty-three, conducted by Olaf Hagen on July 19, 1938. After leaving the Rancho San Francisco and arriving at Santa Barbara, Doty's group of Jayhawkers traded their oxen for horses and continued to the mines. Nine of them went together to the diggings on the Feather River, but when torrential rains washed away their ditches and gravel, Ed Doty and other Jayhawkers set up a camp not far from Oakland to slaughter antelope. "Here antelope were plentiful and hungry travelers, as many as 150 in 1 day, were fed. Later the Jayhawkers that continued in this band went to Santa Clara County [includes San Jose] and put in a crop of wheat. To thresh the crop Doty went to San Francisco and arranged for the purchase of the first thresher that came to California" (1938, 4–5).

On March 5, 1855, Ed Doty (thirty-five) married Mary Ann Robinson (eighteen), who may have come west in a large wagon train that also took the Council Bluffs Road. Henry Doty: "Along the trail wagons of other trains were passed and members of various groups became acquainted with each other. Among those that Doty thus learned to know were the brothers of his future wife, Mary Ann Robinson" (ibid.).[8] Ed and Mary Ann had four children.

In 1878 Ed bought five hundred acres for $3,000 in Las Varas Canyon about "20 miles from [S]anta barbara, west 2 miles from the beach in one of the most lovly spots on earth." It is about four and a half miles north of Goleta, where his brothers Martin and Albert also settled. He called it "Ever Green ranch " (E. Doty 1884). Las Varas Canyon was referred to locally as Doty Canyon because so many Dotys settled there. The Doty Geodetic Survey marker (EW8126) is still on Doty property four and a half miles south of Las Varas Canyon (Arlene L. Doty, July 1, 2015, personal communication).[9] Ed tantalized the Jayhawkers who were attending the reunions in the snowy, blustery Midwest when he wrote, "[Near Santa Barbara] vegetation in places is from 6 to 10 inches high, pinks, verbenes, geraniums are in blossom" (1884), and "i had ripe Black berrys & rasberys on christmas & new years" (1886).

Some two and a half years before Ed died, Manly wrote in 1889: "Ed Doty & wife made us a visit a short time ago & I tell you we had a real good time talking over old times.... he is getting to appear somewhat feeble & looks older tharn myself although we ware both born April 6th 1820. he had his youngest son acidentally killed by a companion while deer hunting. this nearly sets him crazy · my wife & Self are comfortably situated."[10] Captain Ed died June 14, 1891, and was buried in the Protestant cemetery in Santa Barbara (H. Doty 1938; F. Doty 1892). Chester Doty, Edward's grandson, donated the rifle Ed carried through Death Valley in 1849 to the Death Valley National Monument (now a National Park).[11]

Tom Shannon also kept the old rifle he carried through the deserts. He wrote Colton in 1893 that "the only relick[s] I have left that [I] caried across the desert is the old yellow rifle and the little brown handeled Stilletto that you gave me at Devils Gate on Sweetwater."

That same year Manly wrote about Tom and his rifle:

Tom Shannon, an honored Pioneer, residing in the foothills near Los Gatos, is the owner of an interesting relic of the days of '49.... There was no game in this poor part of the world [the California deserts], and many of these early travellers threw away their guns. Indians were rarely seen; but the subject of this sketch held on like grim death to his gun.... Every few days he would be asked to shoot an ox, which was all the food they had. The owners of the oxen agreed to have one killed alternately. Finally it came the turn of Tom's favorite ox to be killed. Tom hated awful bad to kill the loved animal, but it must be, and his gun did it.... Hardly any other rifle of the company was so fortunate as to have an owner with strength and determination sufficient to bring a chunk of iron from Illinois to California weighing sixteen pounds.... The rifle is still the property of the original owner, hanging on curious wood-hooks at his home. (October 15)

In 1900 Tom wrote: "I gave the old yellow gun that I packed across to the pioneer Society as a relic of 49."[12]

Thomas Shannon was twenty-four when he headed to the goldfields, partly to improve his health. He had fought in the Mexican-American War, where he contracted some disease, possibly malaria. While mining in the Feather River–Marysville area, he met and married Amanda Blackfield, and, settling in California, they reared seven children. Tom enlisted in the Union army, was assigned to Tucson, Arizona, and was mustered out in 1866. He moved his family to Los Gatos (southwest of San Jose), where he homesteaded 160 acres. His children helped with the farm, and as his eyesight deteriorated, his daughter Lulu wrote his dictated letters to the Jayhawker reunions. Manly commented in July 1894, "T. Shannon has white hair & beard & looks honestly older tharn a few years ago. *Stevens* is younger & well." The farm supported the Shannon family until 1898 when Tom wrote Colton,

> to let you know that I am still living and that my life has been a financial failure, and I want to ask you if yours has been sucesful enough to advance me two or three hundred dollars to either pay my funerel expences or u[n]til I can pay you back. I have a little home of 50 acres near los gatos that suported me for 30 years but has failed this year.... I can borrow here on mortgage but will not do so. I received the photo of Shelden Young for which thanks.[13]

Since a penciled note reads "ans Sep 27, JB," Colton responded and probably sent some money.

Tom again wrote the reunion from where he was now living in the Soldiers Home in Los Angeles: "I am 77 Jany 25, 1902. am able to walk about with a guide as vision is gone—and eat 3 meals a day. I am kindly attended to here." A year later Tom sent a letter to the reunion saying he could not attend because, "my infirm body and sightless eyes hold me a helpless captive—But I am glad to say that my mental faculty and my memory is extremely good and with thought and heart shall spend with you in recalling the memories for which this day has been set apart by us" (February 1, 1903). Nine months later he was buried near his Los Gatos farm. After Shannon's death on November 13, 1903, John Colton recalled, "Thomas Shannon, a giant in size, past eighty years of age [actually seventy-eight], a veteran of the Mexican and civil wars, has been through extreme hardships; for the past five years totally blind, and for the last seven months confined to his bed with paralysis, but in all of his trials he has borne up with the same stalwart nerve that brought him through the Death Valley" (November 22, 1903; *San Jose Daily Mercury*, November 16, 1903; R. Shannon 1903).

John Colton came west to attend the 1901 reunion held at Dow Stephens's home in the heart of San Jose. Stephens wrote Colton, "Mr. **Manly** is quite feeble but yesterday he was in town the first time for some time past but if the weather should be pleasant we will try and get him here" (1901). Two years later Stephens had more news: "I attended Mr. Manly's funeral yesterday." He died February 5, 1903, the day after the Jayhawker reunion hosted by Juliet Brier at her son John's home in Lodi, California, which Colton and Stephens attended.

Lorenzo Dow Stephens was twenty-two when he left for the goldfields. He found mining exciting, but it did not pay well. Dow bought a team and made a fairly good living hauling timber and other goods, and he tried his hand at farming in Santa Clara County but found he could make more money putting together machinery for others. Several times he went into various partnerships but usually regretted the arrangements. For several years he was a cattle rancher, with his headquarters in the delta country about seventeen miles southwest of Stockton.

Stephens married Julia Ludlum in 1867, and they had two children of whom he was very proud. The Stephens family lived in San Jose, but *settled* is hardly a word that fit Dow. He kept his hand in the mining business. "I am hid away up here in the Siera Nevada Mountains Mining for gold as in the days of '49" (1886). Another year, he arrived home from "the Snowy regions of the Siera Nevadas" to find "Geranium, the Japanica, the fuchia and many other rare flowers all in bloom" (1883). He tested various strikes—Alaska, Aurora-Bodi, and the Comstock—but never found his pot of gold.

Over time Stephens developed a business in San Jose to manufacture and sell windmills. At age eighty-nine he wrote highlights of his adventures in *Life Sketches of a Jayhawker of '49: Actual Experiences of a Pioneer Told by Himself in His Own Way*. Dow was the last surviving Jayhawker to pass "over the river" on February 10, 1921, at age ninety-three.

Edward F. Bartholomew returned to Illinois in 1853 and married Cordelia Kellogg, the sister of another Knox County Company pioneer, William Kellogg (Hopkins 1903, 584).[14] Ed was a corporal in the Civil War. In 1872 he was in Pleasanton, Kansas, where "in my travels I have met with a Sister of our old friend, Wm Rood & Rev Mr. & Mrs. Briar." He partnered with his brother in the grocery business in Buena Vista, Colorado, then worked with a son (one of his four children) in the nursery business in Pueblo, Colorado (E. Bartholomew 1877, 1889; G. Bartholomew 1885, 322). His letterhead announced he was a "Dealer In

And Grower Of Choice Nursery Stock; Forest Seedlings For Timber Culture A
Specialty." Bartholomew wrote numerous letters to the reunions—the last just
thirteen days before his sudden death in Pueblo at age sixty-three, on February
13, 1891 (obituary, February 14, 1891).

George Allen stayed in California and in 1873 homesteaded several mineral
springs in Lake County. He built a Victorian "spa" that advertised its waters as
"successful in curing rheumatism, dyspepsia, dropsy, diseases of women, general debility, and all kinds of skin disease." George never married and may have
developed the springs to improve his own health. He died at age forty-six of an
arterial aneurysm on September 11, 1875 (R. Allen 1877; see also Nelson 1999, 21,
22).

Urban P. Davison (not Davidson) "led a wild & recklous life" and "injoid the
society of allmost every class of people" before he married a young woman in
1875, settled down in Wyoming, and had several children (1878). "I have been
troubled with Heart Dsease for a Good Many years," he wrote Colton in 1883. By
1891 he had "a Family of five Motherless Childron I think is worthy of the kind
treatment of their father." Urban sent Colton his portrait in 1893 when he was
living in Frémont County, Wyoming. He died December 18, 1903, a little more
than a year after he and Colton celebrated the 1902 Jayhawker reunion at Urban's
home in Thermopolis, Wyoming. Urban wrote in 1898, "I met Harison France in
Idaho in the year of sixty three. Being 35 years ago that is the only one since 52
when I met Charles B. Mecum on the Isthmas of pannahma on my last return
from Galesburg as he was returning to Galesburg from California."

Harrison B. Frans wrote on March 22, 1892: "Mining has kept me poore
every Since I hav bin on this coast, but I am Shure now that I will come out all
rite" (see also Nelson 1999, 27–29). He married Mary Ann Brewer October 1,
1853, in California, and they had three daughters (findagrave.com, memorial
#144063536), but he was living alone in Rye Valley, about thirty-five miles south
of Baker, Oregon, in 1870. Ten years later the Census listed him living in a boardinghouse in Baker City and working as a saloon keeper. He wrote in 1895: "I am
making my money the hardest way by pounding it out of rock. I cleand up last
year, settled up and had just five dolars left this year" (U.S. Census 1879, 1880;
Frans 1895). But the following year he owned and was running a quartz mill and
had "ten and twelve men employed running mill & mine. going to make plenty
of money in the near-future." In 1901 he still had "several good quartz claims, and
hope to make another sale in the near future." Frans had that eternal flame of
hope, the hallmark of all prospectors, and loved the life he led: "I can not run as
fast as I could when the Piutes were after me. I have 3 good Rifles and 3 good gray
hounds, And am still raaring and taaring up and down Baar river." Harrison's

son-in-law was a cashier at the First National Bank of Baker City who received Harrison's mail when he was out prospecting (Frans 1900). Frans died January 16, 1902, in Baker City.

John Wesley Plummer was the second-oldest Jayhawker at age forty-three when he traveled to the goldfields. He had been married for twenty-two years, so he probably left most of his nine children at home with his wife when he headed for California. In 1871 the family moved to Toulon, Illinois (about thirty-five miles northeast of Galesburg), where John died on June 22, 1892 (obituary, n.d. 1892).

Tom McGrew died "in or near Boyce Mines [Idaho]. He never returned to the States since he first went to California.... [I] understand that he was a 'Mason' and buried under the auspices of the Masonic Order" (M. McGrew 1872). Colton annotated Tom's photo, saying it was taken from an old daguerreotype from the 1860s and that Tom died in 1864. (The photographer touched up Tom's eyes.)

John Groscup was probably the least successful Jayhawker financially, but he was rich in family and his faith in God. Being born in Bavaria may have contributed to John's delightfully creative spelling and poetic turn of phrase. He lived most of his life in California and in 1872 asked the reunion brotherhood for money to buy sheep (U.S. Census 1870; see also Nelson 1999, 30). The year before, he was married, probably to a widow with several children, because by 1875 he had "a wife and 9 children." Sadly, his wife died three years later, leaving him "with 4 small Children and am keeping house with them" (Ellenbecker 1993, 96; Groscup 1875 and 1880). John was still in Laytonville, Mendocino County, where the 1904 reunion was held at his home. He died February 24, 1916.

Asa Haynes left Stockton September 24, 1850, after seven and a half months in California, to return to his large farm and other businesses south of Knoxville, Illinois.[15] He hosted the Jayhawker reunion in 1885, and the following year he said, "I am an old man, will be eighty-two years old the ninth of February" (1886). He died March 29, 1889, at age eighty-five, and was buried in the Haynes cemetery at DeLong, Illinois, about seven miles south of Knoxville. (In his photo, the hair on Haynes's head was added in pencil, and the ear on the left was also touched up.)

Sidney P. Edgerton (1871) was thirty-eight when he ventured to California. He was in Muscatine, Iowa, in 1872, then seven miles south of Blair, Nebraska, by 1876. He expressed how very much he wanted to attend the early reunions, but never succeeded. He died January 31, 1880, according to Colton's note on Sid's photo, which also adds "copy of an old photo." It is heavily touched up by the copy photographer.

Alex S. Palmer's photo had a Colton notation saying Alex died in Chandlerville, Sierra County (near Slate Creek, branch of the Yuba River), a rich strike when Alex died March 27, 1854. Colton also noted the photo was made from a daguerreotype taken in 1849. If so, it was before Palmer left for California, possibly the only picture the family had of him.

John Leander Woolsey died in Oakland October 8, 1881, age fifty-five. His picture was taken about 1852.

After crossing rugged San Fernando Pass from Rancho San Francisco into the Los Angeles basin, the Brier family spent "ten weeks" in Los Angeles. The **Reverend James Brier** said about their arrival in Los Angeles on "Saturday evening, February 23, 1850":

> We were in absolute poverty, but the gamblers got together and made up a purse of $40, which they gave to me. That was a fortune to a minister in those days. That Sunday I preached in old Judge Nichols's house, on Main street.
>
> I staid in this city ten weeks, and then went to San Jose. Twelve years ago I went back to Death Valley, with Gov. Perkins and party, as a guide, when they were seeking the lost Gunsight mine. I remembered every foot of that awful country; and as I saw the horrid scenes, all the old horror came back to me. (1886)

George Clement Perkins was a state senator from Butte County when he went to the Panamints to hunt the Gunsight lode, guided by Brier. He was governor of California from 1880 to 1883.

After arriving in Los Angeles the Briers had part ownership of a boardinghouse, and for a few weeks Lewis Manly worked for them (Manly 1894, 273–74; James Brier 1886).[16] From there they may have traveled north to San Jose, but they settled in Santa Cruz, where Juliet (or Julia as her family called her) may have helped Abigail Arcan when Abigail's baby was born on July 1, 1850. Julia S. Arcan lived nineteen days and was the first person to be buried in the new Protestant Evergreen Cemetery. Abigail was five months pregnant when, in a starving condition, she walked the 250 miles from Death Valley to the Rancho San Francisco (J. and L. Johnson 2013, 6). Abigail may have named her baby Julia to honor Juliet Brier.

The rigors of **Juliet Brier**'s Death Valley trek stood her in good stead as she and her children shared the life of an itinerant preacher. James Brier was assigned to preach in Marysville (1851), Feather River (1852), Sonoma (1853), Columbia and Sonora (1854), Ione and Volcano (1856), the Santa Clara circuit (1858), and San Francisco (1857 and 1860) (Anthony 1901, 15, 51, 64, 111, 122, 124, 127, 160, 168).[17] During this time Juliet bore three daughters: Mary Caroline (August 15, 1851, in

Santa Cruz), Helen Newcomb (September 25, 1853, in Santa Rosa), and Alice Louise (born 1856[?], possibly in Ione, and died near French Camp in 1857).[18]

James broke with the Methodist Church over his objections to slavery and joined the Congregational ministry. He was on the first board of directors of what is now the University of the Pacific at Stockton.

The Brier family was again in Santa Cruz when the boys were attending Wesleyan College, but in 1879 James wrote the Jayhawkers from Glen Dale Farm, Nevada County, California: "Mrs B & myself are dwelling alone, on our farm in Nevada Co. Our children are scattered. Columbus, the eldest son, resides in Oakland, & teaches in San Francisco. John W. is a Congregational Minister, & is stationed for the present, at Eureka Humbolt Co. He is called 'the Beecher of the Pacific' by many. Kirke White, is teaching in Sacramento, & is Vice Principle of the High School with a salary of $2500.00 per annum. So much for the three lads who were in your Company."

In 1884 James wrote, "Myself and wife are traveling as Evangelists, & we are both quite strong. I lately preached 80 sermons in seven weeks." Finally, by 1893, James and Juliet settled near their preacher son, John, in Lodi:

> I live on my little farm of 3 Acres in the Suburb of Lodi. I do all my own work, & preach frequently, though not in charge of any Church at present. Mrs B——— is still able to do her own work. My eldest son is Prof of Mathematics & Natural Sciences in the Urban Academy San Francisco. My second son is Pastor of the Congregational Church in Lodi & stands at the head, as a Pulpit Orator on this Coast. My wife & I were born in the year 1814 & are therefore in our 79 year. We still hold fast to our Christian faith & the hope of an endless life beyond the deserts of time & Earth.

Kirke White Brier died age thirty-seven on January 14, 1883, employed as a teacher in Sacramento. He left a wife and four children.[19] Christopher, a professor of mathematics and natural sciences in San Francisco in 1893, died in Oakland December 7, 1907. The Reverend James Brier died in Lodi November 2, 1898, age eighty-four. Juliet hosted three Jayhawker reunions in Lodi—1903, 1908, and 1911—and the 1913 reunion was brought to her abode in Santa Cruz about five months before her death May 26, 1913, just shy of her one hundredth birthday.[20] Middle son John Wells followed Juliet in death nine months later on February 23, 1914.

Sheldon Young and **Wolfgang Tauber** mined on the north fork of the Yuba River after arriving in Sacramento about March 6, 1850. By October they were in San Francisco, where they had their pictures taken and booked passage for Panama on their way home. While at sea, Wolfgang died, November 15, 1850.

After crossing the Isthmus of Panama, Sheldon sailed for New York, where he deposited his and Tauber's gold in a bank at eighteen dollars per ounce. He made arrangements to send Tauber's earnings to his family in Germany, and a thank-you letter from Tauber's family, in old German script dated June 1852, is still in Sheldon's descendant David Duerr's possession.

Sheldon was born in Connecticut July 23, 1815, and died at home in Moberly, Missouri, August 18, 1892. After his wife died in 1887, Sheldon chose to live alone. He had a son, Louis Clark, and stepson, Albert Hayden (Young 2010, 55–59).

In 2016 David Duerr, Sheldon Young's great-great-grandson, started looking for Sheldon's grave. In the Oakland Cemetery in Moberly he found a beautiful granite monument for Sheldon's wife, Kerrene, and the marked graves of Sheldon's son and daughter-in-law. But there was only a depression in the ground next to Kerrene's monument—nothing else to indicate the final resting place for the brave Death Valley '49er. David felt this was a disgrace, so in 2016 he made a durable headstone and with the help of the cemetery sexton placed it on Sheldon's grave. "I took a picture and stood in silence for a moment. After one hundred and twenty-four years, my Great, Great, Grandfather, the Sailor, the Pioneer, the '49er, the Jayhawker, the Shipwright and Father, finely had a tombstone to mark his place. RIP" (Duerr 2016, personal communication).

As Manly said about his book, "much is left out," and the same applies here. An appropriate end for this tome comes from an article Lewis Manly wrote about Tom Shannon:

> In closing this short narrative of an old pioneer, I call to mind the stanzas of poet Clifford Trembly:
>> "Toward the west he fixed his gaze,
>> And journeyed on through heat and cold,
>> While always further on a ways
>> The mirage beckoned on as of old.
>> On and on until he found
>> The goal he sought, in treasures vast,
>> For all the mighty earth is round,
>> And he had wandered home at last." (October 15, 1893)[21]

"Constitution and By-Laws of the Knox County Company, Illinois," Plus Roster of Company Names

An article published May 30, 1849, in the Kanesville, Iowa, *Frontier Guardian* includes the constitution, bylaws, and membership roster of the wagon train called the Knox County Company, Illinois. The constitution and bylaws were accepted by the company on May 12 while the wagons were still in northwestern Missouri, heading north to Kanesville. The first roster was signed May 16 (Lorton 1849). However, it did not include the women and children in the company. At that time, most of the men were from Knox and surrounding counties in western Illinois—hence the name. The roster published by the *Frontier Guardian* includes groups that joined the company in Kanesville doubling the original membership.[1]

Constitution and By-Laws of the Knox County Company, Illinois
For a more perfect organization of the above named Company, we mutually bind ourselves to observe in the strictest manner the following Constitution and By-Laws:

Art. 1. The officers of this company shall consist of one Captain, one Lieutenant, four Sergeants and one Clerk, to hold their offices until the arrival of the company at its destination, to be elected by ballot by a vote of a majority of the company.

Art. 2. It shall be the duty of the Captain to take the command of the company and direct its movements, determine the number of men to be detailed on guard, direct as to the time of turning out, herding and yoking cattle, and see that all the provisions of the constitution are strictly complied with.

Art. 3. It shall be the duty of the Lieutenant to perform the duties of the Captain in case of his absence or inability.

Art. 4. It shall be the duty of the 1st Sergeant to keep an alphabetical roll—call the same each morning, and detail the guard and sergeant of the guard for the following night.

Art. 5. It shall be the duty of the clerk to prepare a roll of the company for the First Sergeant and record the proceedings of the company.

Art. 6. A committee of three shall be elected whose duty it shall be to inspect all the teams, waggons and outfits of the company before leaving the Missouri river, and none shall be permitted to start with the company unless found sufficient in these respects. It shall also be their duty to report to the captain any abuse or mismanagement of teams while on the route.

Art. 7. No one shall be considered a member of this company until his name is signed to this Constitution and code of By-Laws.

Art. 8. Other persons may become members of this company by a vote of a majority of the same, and signing the Constitution and By-Laws.

Art. 9. This Constitution and By-Laws may be altered or others substituted in their place by a vote of two-thirds of the company.

By-Laws

Art. 1. No fire-arms shall be discharged within forty rods [220 yards] of camp, except under the directions of the sergeant of the guard.

Art. 2. No loaded guns with caps on shall be permitted in camp except by the directions of the Captain, and never in wagons while traveling.

[*Art. 3. None printed.*]

Art. 4. No spirituous liquors shall be used in the company except as a medicine.

Art. 5. The Captain may call a meeting of the company at any time—and it shall be his duty to do so upon the request of five members of the company.

Art. 6. A horse shall be furnished for the use of the guide by the company—to be ultimately disposed of for the benefit of the company.

Art. 7 For the first willful violation of any of the provisions of the above Constitution and By-Laws by any member of the company, he shall perform extra duty in the discretion of the Captain. For repetition of the same offence, for refusal to obey orders or intoxication, any person may be expelled from the company by a vote of majority of the same.

Art. 8. Any person selling or giving spirituous liquors to an Indian shall be expelled from the company.

Names

The starred names are the men who became Death Valley Jayhawkers. There were about thirty-six men (and one woman) in all; of that number, twenty-four are listed in the Knox County Company roster, and all or most of them first became what I designate as Platte Valley Jayhawkers.

J. L[eander] Woolsey, Knox [County], Ill.*

Nath. Hurlbut (Hulbert), do [ditto]

J[ohn] Grooscup (Groscup), do*

Ubin [Urban] P. Davison, do*

Oren [Orrin] Clark[e], do *

H[arrison] B. Frans, do*

C[harles] B. Mecum, do*

J[ohn] B. Colton, do*

M[arshall] P. Edgerton, do*

N[elson] D. (John?) Morse, do*

J[ames] E[llery] Hale, do

A[lonzo] C. Clay, do*

John Cole, do*

J. W[ilson] Semple (Temple), do

Alex[ander] Ewing, do

R[obert] C. Price, do

E. N[orman] Taylor, do[2]

L. D[ow] Montgomery, do

Cephas Arms, do

John H. [C.?] Ewing, do

Edw. McGowan, do

Asa Haynes, do*

J[ohn] W. Plummer, do*

George Allen, do*

Thos. McGrew do*

John L. West, do*

Edward Doty, do*

Bruen Byram, do*

W[illiam] B. Rude, do*

Aaron Larkin, do*

Alex. Palmer, do*

Tho[ma]s Shannon, do*

Robert Kimble, do
H[enry] J. Ward, Warren [Co.], Ill.[3]
Jno. D. Thompson, do
J. Mackey, do
F[rederick] S]axton] Kellogg, Peoria [Co.], Ill.[4]
L[uzerne] Bartholomew, do[5]
[Edward] F. Bartholomew, do*
Edw. Kellogg, do
Wm. Kellogg, do
Wm. B. Lorton, N.Y. City
S[idney] P. Edgerton, Galesburgh [Ill.] *
D. C. Norton, Knox [Co.], Ill.
J. R. Parker, Fulton [Co.], Ill.[6]
Jacob Grimm, do
J. B. Anderson, do
S. Wise, do
Theodore Ingersol, do
John Chatterton, do
John Lawrence, do
F. B. Shannon, do
H. D. Walker, do
R. Cunningham, do
Orville Jones, do
James Dunn, do
Wm. Whaley, do
J. B. V. Wallace, do
John Short, do
Otho Berkshire, do
R. Baldwin, Erie Co., Pa.
H. Surmeier, Adams [Co.], Ill.
J. B. Surmeier, do
John Lake, do
F. Muer, do
John A. Roth, do
Fred Ketzler, do
John Spees, do
J. D. Campbell, Washington County, Iowa
D. L. Fidler, do
Peter Buck, do

J. H. Cooper, do
Alex Sandilan, do
Geo. W. Buck, do
J. Hasbrook, Rock Island, [Ill.]
S. Shelhammer, do
John Herald, do
Wm. Petre [Petrie?], do
C. W. Miller, Mercer [Co.], Ill.
John Evans, do
L[artes]. F. Langford, do
Thos. Gordon, do
James Merrifield, do
James H. Ellis, [no origin for these three
    entries]
Joseph Ellis,
A. Barton,
Henry Baxter, Capt., Fa[y]ette Rovers of
    Jonesville, Mich.
G. W. Halsted, Lieut. [of the Rovers], do
H[irman] W. Platt, Sec'y. [of the Rovers], do[7]
J[ohn]. S. Lewis, Treas. [of the Rovers], do
C. R. Ralph [or Rhalph, Calvin], do
Adonijah S. Welch, do
[Gustavus] C. Cooley, do
A. J. Baker, do
Ira Latham, do
A[mbrose] M. Dibble, do
Andrew Hartman, do
J. F. Underdonk, M.D., do
P. P. Acker, Kalamazoo, Michigan
G. A. Gale, do
C. L. Cobb, do
Henry Gregory, do
Robert Taylor, Knox [Co.], Ill.
Ira Wells (surgeon of the company), Rock Island
    Co., Ill.[8]
Luke Wells, do
Alex Wells, do

At a meeting of the Company on the 12th of May,—this Constitution and code of By-Laws were adopted and the following persons elected under it, as provided by the Constitution, viz: Asa Haynes, Captain; Cephas Arms, Lieutenant; Thomas Shannon, First Sergeant; Edward Doty, E. N. Taylor and Charles B. Mecum, second, third and fourth sergeants; H. C. Price, Clerk; J. L. West, F. S. Kellogg, and J. R. Parker, Committee of Inspection.

Notes for Appendix A

1. Carl I. Wheat was the first to find this news item from the *Frontier Guardian*, and he published it with a short introduction (Wheat 1940, 103–8; see also Arms in Cumming 1985, 137–42; and Rasmussen 1994, 1:64–65).
2. Mr. Taylor traveled with his wife and young son.

3. Henry Ward, a new friend of Lorton, was in the Lorton-Kellogg mess. Ward had been Edward Kellogg's son-in-law before Ward's wife died a couple years before.

4. Frederick Saxton Kellogg, Edward's cousin, was in Lorton's mess, as were Edward and his son William, who was about Lorton's age.

5. Luzerne (1812–65) and Edward (1828–91) Bartholomew were brothers.

6. The Fulton County men were referred to as the Canton men, a town in Fulton County.

7. First names of the Jonesville, Michigan, Rovers, in brackets, come from Rasmussen 1994, 1:58. The Rovers had been in Kanesville for a couple of weeks before they joined the Knox County Company on May 22.

8. The three Wells men left the Knox County Company to join a smaller company at Elkhorn River a couple days west of the Missouri River.

Names of Death Valley 1849ers

This list is a work in progress. It is here to help the reader; it is not to be considered definitive. It was compiled by Genne Nelson and the author. Genne started with the list of Jayhawkers in the front of Colton's Scrapbook 2, and we added to and checked it by using lists compiled by W. Lewis Manly, L. Dow Stephens, John Ellenbecker, E. I. Edwards, Carl I. Wheat, Burr Belden, the 1849 Knox County Company roster in the *Frontier Guardian*, Colton's wagon list in *Death Valley Magazine*, and references in Jayhawker letters. Included are ninety-five names.[1] Names in parentheses are spellings used in various sources.

The Northern Contingent of Death Valley '49ers
Jayhawker party, Brier family, and others who traveled with or near them for some duration after Death Valley (forty-two persons plus eight who caught up)

Allen, George
Bartholomew, Edward F.
Brier, Christopher Columbus, child
Brier, Rev. James Welsh
Brier, John Wells, child
Brier, Juliet (Julia) Wells
Brier, Kirke White, child
Byram, Burin (Brewen)
Carter, W.
Clarke (Clark), Orrin Charles (Charles Orren?)
Clay, Alonzo [Lon] Cardell
Cole, John Hill
Colton, John Burton
Davison (Davidson), Urban P.
Doty, Edward
Edgerton, Marshall G.
Edgerton, Sidney P.
Fish, ____. (called Father Fish)
Frans, (Franse, Franz) Harrison B.
"Frenchman" (Coverly or Jerome Bonaparte Aldrich??)
Goller (Galler?), John. Manly says he was Dutch.
Gould, (M?) ____. Not named on the Jayhawker reunion lists, but in the late 1890s was thought to be the man from Oskaloosa.

Graff, John, friend of Goller. Not on the reunion lists
Gritzner, Frederick A. (Gretzinger or Kritzner)
Groscup, John
Haynes (Haines), Capt. Asa
Isham, William
Larkin, Aaron
McGrew (McGraw) Thomas
Mecum, Charles Bert [Perry]
Morse, John (N. D.?)
Palmer, Alexander S.
Plummer, John Wesley
Richards, Luther A. (Deacon)
Robinson (Robertson, Roberson), C. William.
Rude (Rood, Roods), William B.
Shannon, Thomas
Stephens, Lorenzo Dow
Tauber, Wolfgang (Wolf, Woolf)
West, John L. (Leveritt, Lew)
Woolsey, J. Leander
Young, Sheldon

Men who caught up with the northern contingent in or after death valley (includes four possible teamsters from the southern contingent)

Abbott, S[amuel] S., Bennett driver
Achison, 2 brothers?

Benson, Anderson. *Not counted.*[2] May have been in Badger Co. with the Bennett family.

Bryant, friend of Benson. Not counted.

Harrison?

Helmer, Silas, Bennett driver

McGinnis,? William. Not counted.[3] Might have been with Young-Tauber mess. Scant proof.

St. John, Lummis. May be A. C. (or C. C.) St. John, from the Badger Company, same as Bennetts. Traveled with Briers from Searles Valley.

_____, Patrick (or Patrick _____?) Traveled with the Briers from Searles Valley.

Smith, Joe? Negro, with Nusbaumer's mess to Travertine Spring. May have left Panamint Valley with Bug Smashers.

Vance, Harry.

Bug Smashers—Mississippi/Georgia Boys and Others (Twenty-One?)

The Bug Smashers were composed predominantly of Georgians and Mississippians, although there were men from other states. The "Pontotoc [Mississippi] and California Exploring Company" was well represented. Twenty-two Bug Smashers were recorded coming into the Mariposa mines on January 29 and February 10, 1850 (Evans 1945, 254, 259). Thus, several names are missing from this list, and I do not guarantee its correctness.

_____, Little West. Negro servant of Jim Martin?

_____, Tom. Negro servant of Jim Martin.

Carr, Fredrick William

"Carr," Joseph (Joe) P. Negro servant .

Charles. Negro servant of Townes.

Coker (Croker), Edward

Crumpton (Compton?) William?

Eshom, V.

Funk, David.

Martin, James (Jim), Missouri

Martin, John D, Texas[4]

Mastin (Maston, Masterton)

Nesbit, William W. N.

Osborn, William Thomas

Smith, Joe? Negro? Possibly from Missouri? From Nusbaumer mess.

Townes, Paschal A. (Towns, Townsend, Townshend?)

Townsend? Mentioned in the secondary literature. Could be corruption of Townes.

Turner, Carr E.

Turner, Sidenham G. B.

Ward, Nat

Woods, James

Southern Contingent (Turned South in Death Valley),

Excluding the Four Teamsters Mentioned Above (Twenty-Four)

Abbott, Harrison, Helmer, and Vance, probable teamsters, left the Bennett and Arcan families at today's Eagle Borax Spring in Death Valley and followed the northern contingent over the Panamint Range. Six men (including St. John and Patrick) caught up with the Jayhawkers and Briers in Searles Valley. They are included in the "Others" list above.[5]

Arcan, Abigail Harriet Ericsen

Arcan, Charles E., young child

Arcan, John Baptiste

Bennett, Asabel (Asahel, Ashal)

Bennett, George, child

Bennett, Martha, young child
Bennett, Melissa, child
Bennett, Sarah Ann Dilley
Culverwell, "Captain" Richard A.
Earhart, Henry
Earhart, Jacob
Earhart, John
Hadapp, W.
Manly, William Lewis
Nusbaumer, Louis
Rogers, John Haney
Schlögel, Anton
Schwab (Schaub), T.
Wade, Almira, child
Wade, Charles Elliott, child
Wade, Harry George, teenager
Wade, Henry
Wade, Mary Reynolds Leach
Wade, Richard Angus, child

Notes for Appendix B

1. Three men, Benson, Bryant, and McGinnis, are listed but not counted because they *may* have been Death Valley '49ers. However, the data are not yet sufficient to include them. The *Frontier Guardian* list of the Knox Country Company roster is not complete—no women or children are listed, and Jayhawkers Luther Richards and William Robinson are also missing. Robinson may have joined the Jayhawkers at Salt Lake City.

2. Anderson Benson told his son, Charles, he was with the Death Valley '49ers. Neither his name nor that of his friend Bryant is mentioned (so far) in the literature. An A. Benson is listed in the Badger Company that included the Bennetts. Benson's recollections reflect information about the Jayhawkers that became public by 1895 after Manly's accounts were published. He is also mentioned by Wheat (1939a, 108) as a possible Death Valley '49er.

3. The only data that McGinnis was a Death Valley '49er, tenuous at best, is a letter from Fredrick Gritzner found in *Deseret News*, August 17, 1850, repeating an article from the *Joliet Signal* that says "Messrs. McGinnis, Young and Wolfdiver had arrived safe with him" in San Francisco. The other three were included as Jayhawkers. Did McGinnis turn back at Mount Misery and connect with the other three in Los Angeles? Other emigrants who turned back reconnected with Death Valley '49ers in Los Angeles, so this is a possibility.

4. The two Martins are differentiated by Ed Coker (in Manly 1894, 373).

5. Brier says Harry Vance was with them in the Slate Range after they crossed Panamint Valley from Post Office Spring. This was several days before the others are reported showing up in Searles Valley.

Jayhawker Reunions

They form a bond of Brotherhood which time cannot efface.

—Charles Mecum 1878

As time distanced John Colton from his escape from the Great American Desert in 1850, he realized, as did other Death Valley '49ers, that his long and rewarding life had hung on a tenuous thread of survival from that near-death desert experience. He needed to remember it; he needed to honor it; he needed to share it. Thus was born Colton's brainchild—the Jayhawker reunions. The *Beaver Valley Tribune* (1898) explained his position:

> Mr. Colton, the originator and perpetuator of the reunion organization, is indefatigable in his search for facts concerning the '49ers and the unwritten history of the west. He has three enormous scrap books filled with every bit of information, incident, illustration, and narration obtainable concerning the '49ers and the history of the pioneer days. He has endeavored to trace the life of each member of the party from the time he was delivered from Death Valley until the present or until death intervened. This task has been an arduous one, and he has succeeded beyond the most sanguine hopes.

A typical reunion invitation, this one from Asa Haynes in 1881, stated the reason for the gathering:

> Dear Sir. The anniversary of the "Jay Hawkers of '49" is approaching and in accordance with the established custom, I cordialy invite you to the reunion at my house on February 4th at 10 ocl. a.m. to renew old friendships & talk over old times when we were lost in the "Great American Desert." We are so widely seperated that I cannot expect all to be there in person and if unable to be there, I earnestly request you to write immediately upon receipt of this a letter to be read at our meeting as you will know how it will rejoice those present to hear from one of the old Crowd.

Only reunions of military comrades in arms can compare to the Death Valley Jayhawker reunions where men gathered together or wrote letters year after year to maintain their fellowship over a span of thirty-eight years until there were no longer Jayhawkers to reunite. The reunions started February 4, 1872, twenty-two years, to the day, after the "walking skeletons" stumbled up to the casa of the Rancho San Francisco near today's Castaic Junction. They continued yearly through 1918 as a celebration of life over death. These reunions were held in the homes of Jayhawkers, mostly in the Midwest, but later some were held in California, one in Thermopolis, Wyoming, and one in eastern Oregon. Even after the deaths of the original Death Valley Jayhawkers, some of their descendants continued to meet sporadically to celebrate the lives and travails of the original '49ers.

From personal reminiscences written for these reunions, saved by Colton and preserved in the Huntington Library, comes much of the material that breathes mortal essence into the living skeletons that crossed hundreds of miles of desert on foot.

Dow Stephens wrote in 1881, "While you are gathered arou[n]d the bountiful Spread Table of our much esteemed host to day, think of where you was thirty two years ago, in the middle of a great Desert and Starvati[o]n Staring you in the face. how pleasant it must be to contrast the present with the past."

At a reunion:

the day was given over to the narration of reminiscences of the horrors of that pilgrimage, anecdotes of the gold diggings, the reading of letters from the absentees, . . . and the reading of extracts from books and papers touching upon the history of the dark days of '49. The voices of the brave '49ers became husky and their eyes dim as they spoke of some comrade who succumbed on the desert ere the party reached the land of plenty; and again, their laughter rang out in hearty peals as some amusing story was told of another's ludicrous predicament. As has been said: "There was pathos in their merriment and joy in their sorrow."

At 2 o'clock the company assembled around the festal board, the richness and profusion of viands contrasting with the stories of rawhide soup, jerk, and blood hash prepared by Mr. Richards, who was camp cook (*Beaver Valley Tribune*, 1894).

The exceptional depth of feeling among these men was expressed by Tom Shannon, who wrote the 1881 reunion: "Some 30 years has passed since we first met, and our imediate companionship was a period of less than one year, but in that short time I believe we formed a brotherly affection for each other that time can never obliterate."

Over time, other Death Valley '49ers, such as Lewis Manly, and friends or family members were invited to attend or write letters to the reunions (Manly in Woodward 1961, 27). After Manly was invited, he diligently wrote letters and sent information to be shared at the reunions.

Edward Bartholomew summed up the value of the reunions: "[They] remind us of the past, not that we desire ever to go through the same again, but to talk over and tell not all, nor the worst, but part of the unwritten history of the lost Jayhawkers of '49. . . . The day we meet to rejoice and mourn. Rejoice that we are yet the spared monuments of His *everlasting* mercy, to mourn the loss of our departed comrades" (1888).

List of Death Valley Jayhawker Reunions

Note: Many of the newspaper articles are found in Scrapbooks 1 and 2 in the Jayhawkers of '49 Collection at the Huntington Library. Newspaper references from 1900 onward are from Palmer's list and have not been confirmed by the author.

Date	Place	Host	Citation
1872	Galesburg, IL	Colton, John B.	Richards 1889[1]
1873	Galesburg, IL	Clay, Alonzo A.[2]	Richards 1889; Doty 1873
1874	Orange, IL[3]	Haynes, Asa	Richards 1889
1875	Near Woodhull, IL[4]	Richards, Luther A.	Bartholomew 1875; Richards 1889
1876	Rippey Station, IA[5]	Mecum, Charles B.	Brier 1876
1877	Galesburg, IL[6]	Colton, John B.	Bartholomew 1877; Colton 1877 JA182 & JA290
1878	Galesburg, IL	Clay, Alonzo C.	Bartholomew 1878
1879	Near Woodhull, IL	Richards, Luther A.	Bartholomew 1879
1880	Rippey Station, IA	Mecum, Charles B.	Bartholomew 1880
1881	Knoxville, IL	Haynes, Asa	Bartholomew 1881
1882	Galesburg, IL	Clay, Alonzo C.	Clay 1882; Richards 1889
1883	Galesburg, IL	Colton, John B.	Colton 1883
1884	Burchard, NE	Richards, Luther A.	Bartholomew 1884
1885	Delong, IL	Haynes, Asa	Bartholomew 1885
1886	Burchard, NE	Richards, Luther A.	Bartholomew 1886
1887	Rippey Station, IA	Mecum, Charles B.	Doty 1887

Date	Place	Host	Citation
1888	Galesburg, IL	Clay, Alonzo C.	Bartholomew 1888; Clay 1888; *Kansas City Times*[7]
1889	Probably Rippey[8] Station, IA	Mecum, Charles B.	Bartholomew 1889; Wiley 1889
1890	Kansas City, MO[9]	Colton, John B.	Colton 1890 JA160 & JA299; *Kansas City Times* Feb. 5
1891	Galesburg, IL	Clay, Alonzo C.	Bartholomew 1891; Clay 1891
1892	Rippey, IA	Mecum, Charles B.	Mecum 1892
1893	Kansas City, MO	Colton, John B.	Colton 1893; *Kansas City Times* Feb. 5
1894	Beaver City, NE	Richards, Luther A.	Richards 1894; *Beaver Valley Tribune* Feb. 5, *Omaha Bee*
1895	Galesburg, IL	Clay, Alonzo C.	*Galesburg Republican-Register* Feb. 5; *Kansas City Journal* Feb. 5
1896	Perry, IA[10]	Mecum, Charles B.	Frans 1896; *The Jefferson Souvenir*
1897	Kansas City, MO[11]	Colton, John B.	West 1897; *Kansas City Times* Feb. 1 and 5
1898	Beaver City, NE	Richards, Luther A.	*Beaver Valley Tribune*, Feb 11
1899	Kansas City, MO[12]	Colton, John B.	Shannon 1899
1900	Richland, OR	Mecum, Charles B.	*Kansas City Journal* Feb. 14
1901	565 Martin Avenue, San Jose, CA	Stephens, Lorenzo D.	Colton 1901; Mecum 1901; *San Francisco Examiner* Feb. 15
1902	Thermopolis, WY	Davison, Urban P.	Stephens 1902; *Big Horn River Pilot* Feb. 12
1903	Lodi, CA	Brier, Mrs. James W.	*San Francisco Chronicle*, Feb. 15; *Out West*, Vol. 18, p. 326
1904	Laytonville, CA[13]	Groscup, John	Groscup 1903; *Ukiah Dispatch Democrat* Feb. 19
1905	Mt. Vernon, IA[14]	Mecum, Charles B.	*Mt. Vernon Hawkeye* Feb. 10
1906	Eddyville, NE[15]	Colton, John B.	*Dawson County [NE] Pioneer* Feb. 17
1907	—	—	*(no information found yet)*
1908	Lodi, CA	Brier, Mrs. James	Mecum 1908; *San Francisco Chronicle* Feb. 9
1909	Kansas City, MO	Colton, John B.	Stephens 1909; *Kansas City Star* Feb. 9
1910	Eddyville, NE	Colton, John B.	Palmer 1910; *Dawson County [NE] Pioneer* Feb. 19
1911	Lodi, CA	Brier, Mrs. James W.	*Sacramento Bee* Feb. 5
1912	Galesburg, IL	Colton, John B.	*Daily Republican-Register* Feb. 5
1913	Santa Cruz, CA?[16]	Juliet Brier at 94 Myrtle St.	*San Francisco Call* Feb. 9
1914	Galesburg, IL	Colton, John B.	*Daily Republican-Register* Feb. 5
1915	San Francisco, Panama-Pacific Exposition	Colton, John B.	Lockley in *Oregon Daily Journal* March 9
1916	Galesburg, IL	Colton, John B.	*Republican-Register*, Feb. 14
1917	Galesburg, IL	Colton, John B.	Mecum 1917; *Republican-Register* Feb. 14
1918	San Jose, CA[17]	Stephens, Lorenzo D.	*San Jose Mercury Herald* Feb. 5

Notes for Appendix C

1. John Colton's Scrapbooks 1 and 2, Jayhawkers of '49 Collection, the Huntington Library (JCHL), contain newspaper articles about most of the reunions. However, the newspaper name is often left off the clipping.

2. Luther Richards wrote Colton the following in 1889: "[The reunion] was at your house in '72, at Lons [Alonzo Clay's] in '73, at Capt Haynes in '74, and at my house in '75, the year Charlie [Mecum] moved to Iowa. I had the dinner at Charlie's house.... We moved to Neb in 1882 and we met at Lon's that year. you look at the 82 paper and see." The Mecums and Richards lived very near one another just out of Woodhull (north of Galesburg) until the Mecums moved in 1875.

3. Orange is a large township in Knox County south of Galesburg, Illinois. Bruin Byram is also listed as coming from Orange. Asa Haynes's property was, and the Haynes cemetery is, located near DeLong in Orange Township, about seven miles south of Knoxville.

4. Woodhull was about twenty miles north of Galesburg.

5. Rippey Station was presumably on the Railroad between today's Perry and Rippey, Iowa.

6. Colton's address is given as Nos. 1 & 2 Public Square, Corner of Main Street, Galesburg, which was his store address. The reunion was probably in his home (return address in Colton 1877).

7. Theodore S. Palmer compiled a list of reunions now in box 9, folder 46, JCHL, 1952, with no JA number. Palmer's list has several mistakes, and the JA numbers used as references are no longer correct (prior to the current collection organization completed in 2000). However, Colton regularly sent Palmer copies of newspaper reports about the reunions, and presumably those references are correct. Palmer also included a list of the Descendants' Reunions from 1928 to 1938 (John Ellenbecker 1993, 111–23).

8. The 1889 reunion was scheduled to be held at Asa Haynes's home, but he was too ill to hold it. Asa died fifty-three days later on March 29. The meeting was presumably held at Charles Mecum's home at or near Perry, Iowa; at least Mecum sent out the invitations (A. Bartholomew [Edward Bartholomew's son] 1889).

9. In 1890 Colton's return address was 327 Sheidley Building, Kansas City, Missouri.

10. Colton brought twenty-two pictures of the thirty-six members of the Jayhawkers to the 1896 reunion. Three Jayhawkers were in attendance and eleven were still living (Scrapbook 2, JCHL).

11. The *Kansas City Times*, 1897, said, "None of the "Jayhawkers of '49 were able to attend the annual reunion.... However, they would have found Colonel John B. Colton, one of the survivors of the expedition, ready to receive them at his home, 439 Bellefontaine Avenue, with a dinner prepared that would have reminded them of what they did not have on February 4, 1850, the day of their deliverance from what seemed certain death from hunger and exposure." Colton assuredly had other guests to celebrate with him.

12. Newspaper articles about the 1898 reunion said the 1899 reunion would be held at Charles Mecum's home. However, Shannon's reply to the 1899 invitation was to John Colton.

13. Laytonville is in Mendocino County. John Groscup (1903) provided rather complicated directions to his house from Willits that included opening and closing four gates and traveling about eighteen miles on the wagon road. He summed up by saying it was "about 5 miles this side of Laytonville whare you turn off to my plase."

14. Mt. Vernon, Iowa, is about 150 miles due east of Perry, Iowa, the Mecums' previous abode.

15. Colton had a large cattle ranch called Buzzard's Roost near Eddyville, Nebraska, about 40 miles northeast of Kearney. Two reunions were held there.

16. At least part of the 1913 reunion was held at 94 Myrtle Street, Santa Cruz, where Mrs. Brier was in residence, possibly while son John was acting as interim pastor or where one of Juliet's daughters lived. Five Jayhawkers were alive—John Colton, Dow Stephens, John Groscup,

Juliet Brier, and John Brier, who was a child in 1849. Groscup could not attend. Colton rarely included the Brier boys in his lists of Jayhawkers still living (*San Francisco Call* 1913).

17. John Colton and Dow Stephens, the remaining living Jayhawkers, celebrated what became the last reunion of the original Death Valley Jayhawkers. Colton died October 13, 1919, and Stephens died sixteen months later.

Some Death Valley Myths
and Inaccuracies Challenged

Death Valley's early history is fraught with startling inaccuracies, some created from active imaginations, some from faulty memories, some from rumor, and some from incomplete research or lack of access to correct information. One of the goals LeRoy and I have striven to achieve has been to set straight as many errors as possible in early Death Valley history, and I have continued that attempt throughout the Jayhawker story. However, some inaccuracies or myths require more space than the text provides, and that is the purpose of this addendum.

Errors began to creep in as soon as the Jayhawkers recovered sufficiently at the Rancho San Francisco to continue their travels. Their stories of a gunsight carved from a piece of "solid" silver propelled prospectors back into the desert to search for the fabled Gunsight Lode. Thus, layers of white man's flotsam were deposited across the desert to mingle with clues that applied to the original '49er trails.[1]

In 1900 Frances Mecum wrote (apparently for her Jayhawker husband, Charles), "The *Democrat* changed my report so much that I did not recognize it.... These reporters get things badly mixed." Lewis Manly concurred, "Many so called wise writers have wrote about us for the past forty years, and have nearly all told fibs" (1891). T. S. Palmer wrote, "From inquiries made among persons now living in the [Death Valley] region, I found only the vaguest and most incoherent information as to the expedition whose disastrous entrance into Death Valley gave it the name it still bears" (February 8, 1894). He later added, "Enclosed is a clipping from the [New York] 'World.'... As a sample of a 'Death Valley lie' it is somewhat above the average as far as inaccuracy & exaggeration are concerned" (April 11, 1894).

Recollections of the Jayhawkers themselves changed over the years—dates became unclear, chronology was no longer important, or memories took on a new reality. Like sticky balls bouncing down a windy street, recollections picked up all sorts of litter and changed dramatically as they rolled through time. For instance, the death of William Robinson (often called Robertson) in the Mojave Desert was told with as many variations as there were tellers. Which is right? There are elements of truth in each.

Newspaper articles about the Jayhawker reunions varied tremendously in their accuracy. One example of incorrect information is in the *Galesburg (IL) Republican Register* from 1895:

> The party came out of the Sierra Nevada [San Gabriel] mountains near where the town of New Hall is situated. They always supposed the name of the Spanish [Mexican] ranchman who received and fed them, to be San Francisco [Francisco López], but in 1890 *The Ventura Free Press* submitted the matter to Frank Thompson, [who] subsequently investigated it....
> He says the Jayhawkers emerged into the Santa Clara valley [Santa Clarita Valley, where the Santa Clara River runs through it] near the ranch known as "San Francisquito" [Rancho San Francisco]. The Spaniard's [Mexican's] name was Jose Salazar [incorrect]. He was the step-father of Don Pgnacio [Ignacio] Del Valle [incorrect] of Cumulos, the home of Romona.[2]

The fallacious information was spread through other newspapers. The truth is Francisco López (uncle of the Rancho San Francisco's majority owner, Jacoba Feliz del Valle Salazar) was in residence as majordomo when the Jayhawkers arrived in the Santa Clarita Valley in early February

1850, not the Mexican, José Salazar, nor any other "owners" (inheritors) of the Rancho. Salazar was not the stepfather of Ignacio because Ignacio's mother was not Salazar's wife, Jacoba. Ignacio was Antonio del Valle's son from a previous marriage. The real Santa Clara Valley is three hundred miles north of Santa Clarita Valley, where the Jayhawkers exited the desert. There were four land grants named Rancho San Fransquito in California, but none of them involved the Jayhawkers.[3] The Rancho Camulos of Ramona fame had not yet been built in 1850.

Children and grandchildren of Jayhawkers inserted additional inaccuracies. Edwin Mecum wrote John Colton his desire to "carry on the memories of those who sixty years ago emerged from Death Valley through San Francisquito Pass. A few months ago I visited the Del Valle Ranch and the Home of Ramona, and Nicholas Del Valle who resides there told me he could remember your party at that time, as he was a young boy, and that he often heard his parents talk about the party" (1910). Two mistakes are in this passage: the Jayhawkers exited the desert via Soledad Pass and Soledad Canyon, not San Francisquito Canyon, and there is no record of a Nicholas del Valle living in 1850 as a descendant of either Antonio, Josefa, or Ignacio del Valle. Nor was it the son of Francisco López, the majordomo in residence who befriended the Jayhawkers (Harline 2008; Northrop 1984, 1987). However, there was an Indian named Nicolaso, age fourteen, listed in the Census as living on the ranch with the Lopézes in 1850 (Newmark and Newmark 1929, 69). He may have been at Camulos in 1910 when Edwin visited and remembered the destitute '49ers coming through or heard stories about them told by his elders. Nicolaso would have been seventy-four in 1910.

Edwin Mecum also wrote Colton that "Rev. J. W. Briar (Our dear old Jayhawker) preached the first Methodist sermon in Los Angeles in 1853, and I believe it was through his efforts that the church [the First Methodist Church in Los Angeles] was organized" (1910). Promptly upon Reverend Brier's arrival in the pueblo on Saturday, February 12, 1850, he preached a sermon at "Judge Nichols's house, on Main street," possibly the day following his arrival (James Brier 1886).

Lewis Manly created one of the most persistent mistakes when he called the Mexican land grant "Rancho San Francisquito." He had returned to Southern California in 1860, and the stage ran down San Francisquito Canyon, the same canyon he and Rogers had used to leave the rancho when they took supplies back to the families stranded in Death Valley in January 1850. The stage crossed the rancho over a six-league portion of the property Jacoba del Valle inherited, the portion she and her second husband, José Salazar, had mortgaged by 1860 (Perkins 1957, 19; Smith 1977, 98, 128; book 3 of Mortgages, 469, Office of County Recorder of Los Angeles). That mortgaged portion, a small part of the rancho, was sometimes referred to as San Francisquito in the 1860s. However, the Rancho San Francisco was not yet divided when the pioneers entered it from the desert in 1850, and it had been known since before 1804 and legally referred to as the Rancho San Francisco. After Manly's book came out, several Jayhawkers began using the diminutive name, thinking it must be the correct one, and some recent writers have continued that error.

Another inaccuracy came from a letter Tom Shannon wrote to a reunion, saying Wolfgang Tauber died "on the voyage somewhere in the vicinity of Cape Horn" (October 1, 1897). Sheldon Young's diary makes clear the two of them were heading for the Isthmus of Panama when Wolfgang died and was buried at sea. They were not voyaging around South America via Cape Horn. Young crossed the isthmus carrying with him Tauber's earnings from California, which he then sent to Tauber's family in Germany (Young, December 2, 10, 12, 1850; Duerr 2016 personal communication).

When Lew West read in an 1894 San Francisco newspaper that he was dead, he wrote Colton expressing the contrary (*Daily Republican-Register* 1895). West lived another four years and died in the Sacramento hospital on January 12, 1898 (*Record-Union*, April 1898).

Jayhawker misunderstandings started the confusion about the eleven men who became the "Pinney-Savage" party that left Mount Misery to follow the O. K. Smith packers. Some Jayhawkers confused the Bug Smashers (who packed their backs and separated from the Jayhawkers and Briers in northern Panamint Valley) with the Pinney-Savage party, who also backpacked their meager rations and headed west toward Walker Pass. But the Pinney-Savage party started their solitary march from the Sheep Range in Nevada, 145 air miles east of Panamint Valley, the departure point for the Bug Smashers. Pinney and Savage reached the goldfields, but to date the fates of the other nine are unknown. Searching for them is now possible since William Lorton provided their names: "Chas. McDermet, Jno. Adams, Willey Webster, two Wares, Baker, Semore, Allen, and Mr. Moore (Lorton, November [19], 1849, and January 20, 1850). But that is another project.

Charles Mecum wrote to the first Jayhawker reunion in 1872, "Of those boys [the Bug Smashers] that left us at Silver Mountain [the Panamints west of Death Valley] we are disposed to draw the veil. their Sufferings were to sickening to dwell upon. I saw Savage after I got to the mines and his eyes would fill with tears at the mention of that sad time" (see also Manly 1894, 113–14). Savage was with the Pinney-Savage party, not the Death Valley Bug Smashers. As far as I can determine, the Bug Smashers all got through to the San Joaquin Valley and arrived at the Mariposa mines (Evans 1945, 254, 259). Jacob Stover also saw Pinney and Savage in the gold camp of Nevada City (Stover in L. Hafen and A. Hafen 1954, 274–91). Stover had traveled with two men from Iowa who joined what we call the Pinney-Savage party and he was with them (and William Lorton) when the party left the other packers east of the Sheep Range in southeastern Nevada.

Manly confused the Jayhawkers with the Bug Smashers who traveled in proximity through most of today's Nevada. He wrote Alonzo Clay, "I dont see anything of James Martin & his darkey [who were Bug Smashers] in your list of 38 [Jayhawkers]. If I remember right he was the Captain of your organization & Mr. Brier was not" (1888).

Historians also added inaccuracies. Faulty information professes the two Bartholomews who came west in '49 were father and son or possibly father and two sons. This myth originated in the 1908 *Death Valley Magazine* article that notes occupants of wagon "No. nine, [were] Edward F. Bartholomew [sic], father and brother, of Farmington [Illinois]." Historian John Ellenbecker compounded the myth. The father's name was Noyce, as was one of Edward's older brothers. Luzerne, the brother who was with Edward as far as Salt Lake City, may have seemed to be a father, since he was sixteen years older than Edward (De Laney 1908, 101; Ellenbecker 1993, 88, 99; Rasmussen 1994, 1:65).[4]

There is a Death Valley story, now known to be fictitious, that a Jayhawker (or one of their traveling companions) broke a leg while "camped at Surveyors' Well [north of the Death Valley sand dunes].... When the Indians visited the abandoned camp they found only one dead man... who had broken his leg and had been shot through the forehead and body." This was purportedly a quote from Tom Wilson, a Shoshone Indian, who "consented to tell [Dan Coolidge] the real story" (Coolidge 1937, 18).[5]

This fanciful mercy killing received a new twist when Carl I. Wheat (1939b, 18–19) wrote his fictitious fable "Hungry Bill Talks," a story Death Valley historians and trail buffs have faithfully believed since it was written by a preeminent historian. However, Death Valley history expert Richard Lingenfelter had reservations about the story and later wrote a cautionary note: "Carl I. Wheat also wrote of a purported interview with Hungry Bill...but it is at least partially fictitious" (1986, 45, 485n20). As it turns out, Wheat's entire story was fictitious, but that was not generally known until LeRoy Johnson found a letter from John Beck to Wheat in the Wheat Collection at the Bancroft Library, documenting that "Hungry Bill Talks" was indeed a work of fiction (Beck 1939).

In 1938 historian John Ellenbecker started the rumor that Nevada's first governor, Henry Blasdel, came upon the Brier possessions in Death Valley, where he found, and later returned, Mrs. Brier's silverware. Ellenbecker gives no references in *Jayhawkers of Death Valley* (1993, 35, 37). It is possible he was extrapolating from John Brier's two accounts of abandoning their wagons at Cane Spring in Nevada: "the forsaking of nearly every treasured thing, the packing of oxen, the melancholy departure" (1903), and "discarded comforts and treasures...were thrown out upon the sand, at the base of a wind-beaten hill where, ten years thereafter, Governor Blaisdel found them, still in a state of remarkable preservation" (1911a).

Blasdel's expedition did not go to Cane Spring, where the Briers discarded everything they could not pack on their oxen. John Brier, age six at the time, confused the Cane Spring abandonment site with the Jayhawker abandonment site in Death Valley, and Ellenbecker made the wrong connection. A careful reading of Richard Stretch's report (1867, 145, 147) makes clear the abandonment site the Blasdel expedition visited was Jayhawker Well in Death Valley. From there the expedition traveled to Travertine Springs, east through Ash Meadows thence to today's Indian Springs, Nevada, but they did not go to Cane Spring. On their return trek they camped at White Rock Spring in northwest Yucca Flat, not at Cane Spring, and they make no mention of finding any '49er equipage. Another example of the confusions and mystique that lace Death Valley history.

Carl I. Wheat (1939a, 31n17) started the idea that some of the Bug Smashers—the Georgians— cut due west across the Amargosa Desert from the mouth of Fortymile Canyon and came through the Funeral Mountains in the vicinity of Indian Pass. Koenig, Levy, and Southworth repeated this possibility. Wheat seems to base his assumption on Manly not seeing the Georgians on his hike to Jayhawker Well from Travertine Springs and back to the springs. That's because the Georgians and all the other Bug Smashers had already arrived at the well. Manly makes clear that while at the well he encountered only Doty's mess: "I came to a campfire soon after dark at which E. Doty and mess were camped. As I was better acquainted I camped with them . . . and before daylight in the morning I was headed back" (1894, 140–41). After Manly headed back to Travertine Springs, the Bug Smashers left Jayhawker Well to reach the snowy pass in the mountains (Manly 1894, 255). They left December 29, according to Sheldon Young (see also Young, December 4, 26, 28, 1849).

Wheat's thesis does not take the timing into account, which is complicated to explain. Suffice it to say Wheat's suggested route is half as far (roughly forty miles) as the route taken by the Jayhawkers, the other Bug Smashers, and the Brier family (roughly eighty miles), and the Georgia boys were not encumbered by wagons as were the Jayhawkers, who also lingered beside the Amargosa River to shoe their oxen. Since the distance was shorter and they did not have wagons and did not linger as the Jayhawkers did, had the Georgians taken Wheat's route, they would have reached *Tugumu* (Jayhawker Well) several days *before* the Jayhawkers. Instead, the Georgia boys arrived *after* the Jayhawkers.

These are samples of small inaccuracies or myths that have been perpetuated over the years, but history is more than "a fable agreed upon," as Napoléon Bonaparte suggested, especially when the facts are every bit as inspiring (or titillating) as is fiction.

As time goes on, more information comes to light, but also people interested in the '49ers begin to dabble in material available in print or on the Internet and perpetuate or create new inaccuracies. Or they hike a desert canyon and conclude it must be the one used by the '49ers. Since many canyons fit the same description, the puzzle becomes which canyon *best* fits the whole context of the emigrants' movements through the desert. Only by spending a great deal of time on the ground checking all the possible trail choices and by reading as much of the original (not just secondhand) material as possible can a researcher legitimately come to as accurate a conclusion as possible. Even then, how the material is interpreted is crucial.

After Margaret Long wrote her excellent book about the '49ers, *The Shadow of the Arrow*, in 1941, she returned to the desert area and talked to prospectors, old-timers, and ranchers. Her enlarged and revised 1950 edition is, in our opinion, not an improvement. Her initial intuition and research were not cluttered with the assumptions of others who thought or pretended to know details about the Death Valley '49ers.

Notes for Appendix D

1. For instance, Manly said, "Lew West went immediately back to locate the Silver mine but could not find it & came near being hung for his fooling his new friends" (March 16, 1890; see also "Route of the Silver Mountain Expedition in the fall of 1850" on Gibbes's 1852 map; and Wheat 1939c, 197).

2. Rips in the paper caused some missing sections. Article repeated from Thompson, May 2, 1890, *Ventura (CA) Free Press*.

3. Of the four Mexican land grants named Rancho San Francisquito, one was in Los Angeles

County, one in Monterey County, and two in Santa Clara County (Shumway 1988, 33, 51–52, 85, 98). The Los Angeles County Rancho San Francisquito, adjacent to the San Gabriel River, was granted to Henry Dalton in 1845. See chapter 9 for more information on the Rancho San Francisco and its owners.

4. The family genealogical record mentions the two brothers who went to California in 1849, but it makes no mention of the father doing so (G. Bartholomew 1885, 211–12, 319–20, 321–22).

5. Surveyors Well is about thirteen miles northwest of Jayhawker Well. Jayhawker Well is adjacent to today's McLean Spring, head of Salt Creek.

Data to Analyze the '49er Route(s) from
Mount Misery, Utah, to Coal Valley, Nevada

Below are data (plus what you find in chapter 5) to analyze the Death Valley '49ers' route(s) from Mount Misery in southwestern Utah to Coal Valley in southeastern Nevada, an air distance of about seventy-three miles and a trail distance of about a hundred miles. In addition to the information provided by the '49ers (Sheldon Young's diary, Manly, Nusbaumer, the Briers, Groscup, Mecum), of great value are William H. Dame's "Journal of the Southern Exploring Company for the Desert" and James Martineau's "A Tragedy of the Desert" from 1858. That year Dame and Martineau took Asahel Bennett, one of the Death Valley '49ers, as their guide to explore the country west of Enterprise, Utah, into the White River Valley in Nevada (see L. Johnson 2005).

The Southern Exploring Company route, the same or similar to the 1849 Bennett Train route, can be followed westward from the White Rocks area (headwater of Nephi Draw about five miles north of Mount Misery) to White River wash. (The company did not go south from the head of Nephi Draw to Mount Misery. They continued west from the head of Nephi Draw.)

Asahel Bennett guided the Mormon Southern Exploring Company through part of the geographic Great Basin in 1858 to search for a sanctuary where Mormons could resettle if U.S. troops invaded Salt Lake City in the so-called Utah War (Stott 1984, 20; see also Bancroft 1889, 512–42).[1] The Mormon explorers followed Bennett's 1849 trail into White River Wash.

William H. Dame, leader of the company, was a colonel in the Mormon militia who wielded considerable power in the church's ecclesiastical structure. He was also deeply involved in the 1857 Mountain Meadows Massacre—the slaughter of almost one hundred men, women, and children (above the age of six), members of the Fancher Wagon Train (Bagley 2002, 386–90). James Martineau was the historian and surveyor for the exploring company and drew two maps delineating the expedition's route.[2] He recollected after thirty, and again after fifty, years that they used Bennett's Trail "over one hundred miles" and "about one hundred and twenty miles" (*Santiago* [Martineau's pen name] 1890, 297; Martineau 1928, 271). Some of the explorers broke their code of silence, and Bennett learned the horrid details of the massacre.[3]

A word of caution: the 1849 emigrants and the 1858 explorers had very different goals. We don't know how much the 1858 exploring party deviated from Bennett's 1849 trail to satisfy their charge. For instance, the location the explorers joined Bennett's trail is debatable (is it where the journals say it is near today's site of Heber or where the '49er wagons left the Old Spanish Trail?). Does the expedition route represent the Bennett Train route only (and how closely) or does it include part of the Jayhawker or Bug Smasher route(s), which seem to differ. It does not include traveling north at the beginning of the trek as Manly described the way Martin led them and as delineated on Manly's "Palmer" map. Thus, care must be taken when trying to extrapolate the 1849 Bennett (and certainly the Jayhawker) trails from the 1858 travel diaries. The expedition journals are clearly presented in LeRoy Johnson's "Bennett's Trail Into the Great Basin" (2005, 43–74).

Here is a brief description of the 1857 expedition's route from White Rocks (head of Nephi Draw) to White River Valley. Quotes come from Martineau's "History of the Mission to the Desert" (Martineau, May 3, 1858; also in L. Johnson 2005).

From the White Rocks area (north of Pine Park Bench and west of Nephi Draw), they went down the western flank of the hills north of Pine Mountain to Clover Creek (their Beaver Creek) near today's Brown, a siding on the Union Pacific Railroad, thence down Clover Valley to the impassable sheer walls of its lower end (almost four miles west of today's Barclay where several railroad tunnels pierce this barrier). In 1849, as the wagons broke a trail, Lewis Manly was out scouting from every possible mountaintop. The Bennett Trail then turned north up Islen Canyon, passed between piñon- and juniper-covered Mosey and Empy Mountains in the Cedar Range, and turned west into Meadow Valley south of today's Panaca, Nevada. The trail ascended the Chief Range (southwest of Panaca) to the springs now called Bennett Springs, in honor of '49er Asahel Bennett. From Meadow Valley, Martineau said the 1858 exploration "moved 11 miles to Bennett's Springs and encamped. Here is excellent feed, bunch grass, and two large springs—slightly warm. They are situated near the summit of the divide."

On May 4, 1858, Martineau records that they "traveled to top of divide 4½ miles" to Bennett Pass between the Chief and Headland Ranges, "thence to head of a kanyon 1½ miles, thence down the kanyon (a crooked, narrow, dry wash) to its entrance into the valley 9 miles, thence across the valley [westward in Dry Lake Valley], . . . 13 miles to Rocky Point Hill, and encamped without water, and grass very scarce. Total distance 28 miles. All desert to day. Road sandy & heavy."

The next day the Mormon exploring party traveled north six miles in Dry Lake Valley, where their scouts found a little water in a dry wash. They spent May 5 filling their barrels with water and "found an Indian, whom we treated kindly and gave him bread enough for several men. This he ate at one meal, and then went to his wickiup and brought and roasted his own store of provisions. He roasted and ate two large mountain rats, one mouse (raw), 5 large lizards, one horned toad, and 4 large rattlesnakes, and seemed to wish for more." They called this water source Desert Springs Wells, today's Coyote Spring in Dry Lake Valley.

May 6 they "passed over the divide [North Nopah Range] about 2 miles from the Desert Springs Wells, (as we called our last camp) thence several miles down a kanyon to a long narrow valley [White River], nearly level. Our course here turns nearly north. Camped without water. Distance 27 m."

At this point, to get around the looming Seaman Range, the '49ers turned either south or north in White River wash. I provide one more day of expedition information because their records do not indicate when they leave Bennett's Trail.

On May 7 the expedition continued north in White River wash.

About 5 miles found a little alkali water standing in the bed of a small ravine with new grass growing along its edges. Camped at 1 o'clock p.m. on the border of a large wire grass meadow or slough full of strong alkali water [today's Murphy Meadows]. There is considerable water running here, at its narrowest place the stream is about 20 feet wide and a foot deep, with a good current, but it soon runs to nothing below, forming those little, isolated pools we first came to. Above the camp a short distance the water expands in large, shallow ponds, which were much frequented by Indians [today's Wildlife Area, White River Valley, Nevada]. Traveled to day 11 miles.

Sheldon Young commenced writing his diary at Mount Misery after a two-week hiatus. I present the section between Mount Misery and Coal Valley playa without comment to eliminate bias for future researchers. However, note that some distances have no direction, and some travel has no mileages.

[Nov. 9] [We are waiting at Mount Misery] for some of our men that were out looking for a road · nothing of note occured in camp this Day
[Nov.] 10) lay in camp
[Nov.] 11) this Day left Camp · made a westerly course went 25m · found grass & water · passed over some Broken country · we have verry pleasant weather · cool nights
[Nov.] 12) lay in camp · two horses taken · sent out 8 men to look out a road · had a pleasant Day

[Nov.] 13) Bore a westerly Course · Campt on the side of the mountain at a warm
spring · went 12m had good roads plenty wood & grass · rained

[Nov.] 14) Bore N of W went 10m stony road · had a Dry camp · neither grass or water · it [rained?]
& snowed in the afternoon enough to whiten the ground But went off as fast as it fell · N of W

[Nov.] 15) had good roads · went 6m found water not much grass · it is a dry sandy
Desert · went 8 miles & campt on the opposite side of the valley · no water · grass scarce no
water nor wood · NW

[Nov.] 16) had a hard road over a range of mountains · no grass nor water · it is a Dismal
looking country · there is nothing growing but greasewood · went 8m campt at a
spring · not much grass. Bore N of W

[Nov.] 17) this Day had rough roads for 4m then came into a narrow vally · there is no
running water · rain water standing in pudle[s] · grass is scarce wood there is none · it is a
Dubious looking country · went 10m · NW course

[Nov.] 18) this Day had good roads · went 15m · grass & water scarce · following a narrow
vally · Bearing N.W course

[Nov.] 19) this Day went 5m · came to plenty of grass & water · continuing our cours in the
same vally · no timber · had the best grass since we left the Platt

[Nov.] 20) this Day we left [Barney] wards Muddy & Bore off a S.W course & struck another
vally · went 16m · had good roads & a Dry camp · not much grass

[Nov.] 21) went two miles found water · had good roads · cloudy · some rain · mountains
covered with Snow · in the evening snowed · went 16m had water grass · W

[Nov.] 22) it Snowed verry fast · snow lay on the ground 2 in at 9 oclock · went 4m &
campt · there was snow squalls all day

[Nov.] 23) had a verry cold night · left camp at 8 oclock · went to misery lake · Distance 2
m⁴ · had no grass · had good roads · bore S.W Direction · two ho[r]ses taken

[Nov.] 24) this Day lay in camp · the most of the Day exploring

[Nov.] 25) this [day] went 12m S.W course · had a Dry camp · it looks rather Dubious for
water · misery lake is highly charged with alkali⁵

Unfortunately, Manly provides conflicting information about how long the '49ers traveled
north, and he mentions varying days when the single men caught up to the Bennett Train after
it turned west. Below are the variations found in Manly's information that add confusion when
attempting to analyze the route(s) (emphasis added):

- "So they hitched [at Mount Misery] and *turned nearly due north*" ("From Vermont," December
  1887).
- "He [Jim Martin, chosen captain] contended that we would have to *travel some distance north
  before turning west*" (ibid.).
- "Under Martin's orders *we drove north three days or more* on very high ground" (ibid.).
- "The route led *at first directly to the north*" (1894, 113).
- "[Manly asked] them why they were steering so nearly due north.... They replied that *their
  map told them to go north a day or more* and then they would find the route as represented" (1894,
  330).
- "I soon became satisfied that *going north was not taking us in the direction we ought to go....* They
  insisted they were following the directions of Williams, the mountaineer, and *they had not yet
  got as far north as he indicated*" (1894, 114).
- "Bennett and some others concluded to take my advise and turn west. *The first night Martin &
  Company overtook us*" ("From Vermont" 1887).
- "*The second night* the brave Jayhawkers who had been so firm in going north hove in sight in
  our rear" (1894, 115). [Was this the second night after Bennett's train turned west or, if read
  chronologically as he wrote it, the second night after the pillars of rock at Meadow Valley? Was

Manly differentiating between the Jayhawkers and the Bug Smashers, or are the Jayhawkers here the same group as Martin and company?]

- *"About the third night* the Jayhawkers were overtaken by seven more wagons owned by A. Bennett and friends, J. B. Arcane and family, two men named Earhart and a son of one of them, and one or two other wagons" (1894, 329).
- "Bennett's little train turned west from this point and the Jayhawkers went on north, but *before night they changed their minds and came following on after Bennett whom* they overtook and passed, again taking the lead" (1894, 330).

Asa Haynes's brief log is of little value because he provides no description and rarely provides a clue to where he was. All Nusbaumer wrote about the trail was, "We again turned our faces toward the west after chopping our way through a forest an entire day" (1967, 34). John Brier mentions, "When the train was once more set in motion [from Mount Misery], ax-men led the way.... A rough and hazardous trek was exposed, to follow which tested to the utmost the discipline of the oxen and the will of the drivers" (1911a, 2).[6]

The other quotes—from Brier, Groscup, and Mecum—are in chapter 5, as is an overview of the route.

Notes for Appendix E

1. Whether they were in or out of the Great Basin depends on whether you speak physiographically (yes, they were in the Great Basin) or hydrographically (yes and no). The Meadow Valley drainage (Panaca) is outside the basin (it drains to the Pacific Ocean via Muddy and Colorado rivers), and the White River/Pahranagat drainage is in or out depending on which author you read. Grayson (1993, fig. 2.2) and Francaviglia (2005, fig. 1.1) say it is in the basin, but Karl Musser's map (fig. 2 in Shallat 2013) says it is out of the basin and is drained also via the Muddy and Colorado rivers.

2. The expedition's first map was drawn near Panaca Spring at the end of the expedition and is dated "Meadow Valley, June 24, 1858." The second map is a larger scale, but is undated and no scale is given (Wheat 1959, 4, facing 128; Martineau Map 1858).

3. Manly tells us that after 1858, "Mr. Bennett, [now] being a Mormon, learned so much of their bloody work,...that he could not conscientiously remain and be known as one of this disreputable community,...so he suddenly made up his mind to get to a more free country" (Manly in Woodward 1961, 38–39). Bennett's first name has been variously spelled "Asa," "Asael," and "Asabel," but "Asahel" is correct.

4. In the original diary, Young's number is easily interpreted as "26m" instead of "2m, " but by using twenty-six miles, the mileages are skewed from Young's "Misery Lake" to the Amargosa River. I therefore accede to the "two miles" in Colton's typescript of the diary (Young 1849b).

5. This latter comment about the lake seems to be retrospective to November 23.

6. John was a boy of six in late 1849 and had become a respected and effusive minister by the time he wrote "Argonauts of Death Valley."

Sources: Background

The main sources for Jayhawker material between Illinois and southwestern Utah are the original journals of William B. Lorton in the Bancroft Library; the log of the San Joaquin Company on the Southern Route from Hobble Creek to Mount Misery, now attributed to Adonijah Strong Welch; and John Colton's trail letters to his family.

William Lorton traveled in the same train as the Knox County boys (many of whom became Platte Valley, then Death Valley Jayhawkers), and his California journals relate the company's experiences from western Illinois to Salt Lake City via the Council Bluffs Road (north side of the Platte River), thence south to Mount Misery in southwestern Utah. His journals bring color and detail to the first six months of the Jayhawker trek.

Lorton, a twenty-year-old from New York City, was visiting cousins in Knox County, western Illinois, when gold fever broke out in late 1848. He had graduated from school and plied the trade of house painter before embarking on his journey to see the West (Illinois and Iowa back then). He was also a trained singer and violinist and was noted for his repertoire of Negro songs. He decided to extend his trip to include the goldfields of California, so he bought his share of victuals and two trained oxen and joined a wagon mess. He also volunteered to be a correspondent for the *New York Sun*. With that goal in mind, he kept a detailed journal from which he wrote periodic reports to the *Sun*. His journal ends abruptly in Los Angeles, but we know he managed to get to the goldfields and then ultimately returned to New York, probably with a stash of gold, where he became a clock maker, married Mary Walker Briggs, and fathered three children.[1] He died in New York City in 1892.

The most valuable references for the rest of the trip—from Mount Misery to gold country—are Sheldon Young's diary, Manly's writings, Jayhawker reunion letters, and James Briers's 1886 talk written up in the *Los Angeles Times*.

Sheldon Young's diary, starting from Joliet, Illinois, on March 18, 1849, to December 15, 1850, ending after he crossed the Isthmus of Panama, has been known for years in the form of a typescript that Jayhawker John Colton had made in 1897, but when David Duerr provided us with the electronic scans of the original pages in 2011, a number of errors in the Colton typescript became obvious. Young's diary is the only extant almost day-to-day diary of the Jayhawkers' trek through the Nevada-California desert part of their trip. Although Sheldon Young was not with the Jayhawkers until after Salt Lake City (maybe after Mount Misery), they listed him as one of their own, even though for years they could not remember his name.

David Duerr, Sheldon's great-great-grandson, made electronic scans from the original pages of the handwritten diary before he and his sister, Gretchen Forsyth, donated it to the Huntington Library in San Marino, California. When I quote sections now missing from the original diary (but that did exist in 1897), I use John Colton's 1897 typescript, JA555 in the Jayhawkers of '49 Collection at the Huntington Library.[2] Colton's discovery was explained in a newspaper report of the 1898 reunion: "The diary was written in pencil on the pages of a cheap note book, and the ravages of time had nearly effaced the writing. With the assistance of powerful magnifying glasses, a large portion of the record was deciphered, and then transcribed and bound in durable form" (*Beaver Valley (NE) Tribune*, February 11, 1898). LeRoy and I knew there were problems with the copy John Colton had made, and it was a dream of ours that the original Young diary might someday come to light.

And it happened! The way David Duerr found the diary is the proverbial fairy tale—the dusty shoebox hidden in a dark corner on the highest shelf of a cabinet that required a stepladder to reach. When David was cleaning out the house his mother had lived in for many years, his sister mentioned to him to keep an eye out for an old diary. Sure enough, in his father's den the last room he needed to box up, David climbed a ladder to remove the last of the books, and there he found a dusty shoebox that contained a couple old notebooks, loose, yellowing, handwritten pages, some coins, and papers from the family of Wolfgang Tauber, Sheldon Young's German Gold Rush friend.

The diary pages were disorganized and difficult to read, so David put the box aside. He jeeped across the country a couple of times with the shoebox as part of his baggage, and on one trip he stopped at a gas station near Trona, California, where he happened to buy Burr Belden's *Goodbye Death Valley: The 1849 Jayhawker Escape!* Suddenly, David realized those old notebook pages he was lugging around had something to do with that awful trek in 1849! Back in Florida he scanned the pages into his computer, and the result was his publication in 2010 of *The Lost Jayhawker Story: Sheldon Young with the Death Valley '49ers.*

The following year, David and Gretchen hand-carried the diary to the Huntington Library to donate it for posterity and the use of researchers. They then attended the Death Valley '49ers encampment that honors those intrepid pioneers, which included their great-great-grandfather Jayhawker Sheldon Young.

Young's diary was composed of two paperbound booklets, the first about 4.8 inches wide by 7.2 inches tall with ledger lines printed on each page. It is in very good condition with clear and legible text. But only two pages have script on them, both written in ink, recording only the first three days of Young's trip. The other pages are blank. The second notebook is much smaller—3 inches wide by 4.8 inches tall—and well used. Some of the pages are still in signature form with a small amount of stitching still attached. But most of the pages were loose and misarranged when David Duerr found them.[3] This journal is written in pencil and has sixty-two pages of the trek to California. Two additional pages list ingredients and dosage of a medicinal brew.

Although the Young diary appears to be written daily, there are gaps—at least three gaps of a week or more—not just from missing pages but from Sheldon either misplacing the booklets or choosing not to write in them. His mileages are not foolproof; some days he leaves them out; sometimes he adds mileages that can be proven to be incorrect. Sometimes several entries are clearly written at one time instead of daily, and this is where most of the inaccuracies happen. The diary entries of most value for this project are those from November 9, 1849, at Mount Misery in Utah to the lifesaving Rancho San Francisco in Southern California on February 4, 1850. Most of Sheldon's sentences are fragments—a sensible way to briefly and quickly record information in a diary.

Two other extant diaries of lesser value were also kept during the trek—that of Asa Haynes, elected captain, then colonel of the Knox County Company, and that of Louis Nusbaumer, a German-speaking gold seeker from New York who turned south in Death Valley. Haynes's diary and its rewrite are also in the Jayhawkers of '49 Collection at the Huntington Library (JA1051). Haynes's recollections were written down by his daughter Nancy Haynes Wiley over a period of years and later organized and rewritten by her granddaughters in a jumble of vignettes. Haynes's diary is a log of distances with few cardinal directions and almost no commentary; I use his mileages only when he gives a location that is recognizable.

Nusbaumer's diary, plus the 1933 translation made by his daughter, Bertha Whitmore, plus a typed copy by Max Knight, are in the Bancroft Library in Berkeley, California. Bertha's translation, edited by George Koenig, was published as a Bancroft Library Keepsake, *Valley of Salt, Memories of Wine*, and a new translation was made of the Death Valley portion for our *Escape from Death Valley* (1987, 158, 160–68).

The journals of Lorton and Welch flesh out the trek from Salt Lake City to Mount Misery.

Welch was chosen to help draft the San Joaquin Company's constitution and bylaws, which were adopted on October 1, 1849, by the (male) train members, and four days later Welch was appointed scribe "for keeping the journal of the Route." Welch's journal covers September 30 to October 14 and October 20 to November 5, 1849, from Hobble Creek (about ten miles south of Provo) to Mount Misery on the border of Utah and Nevada.

Originally considered David Switzer's Log, historian Dale Morgan determined it was written by Adonijah Welch, who at some point after reaching Mount Misery gave his notebook to David Switzer. Theodore Ressler transcribed it as "The Switzer Log" in his mimeographed *Trails Divided*. Ressler thanked "Geo. D. Switzer of Beaumont, Texas, for supplying [a] copy of the journal of the 'San Joaquin Company,' handed down by David Switzer." Because Ressler had permission to publish the log from the owners, we use his book when quoting Welch. A photocopy of the original log is in the Bancroft Library, but they do not own the original. Welch also wrote five letters based on his personal journal (still undiscovered) that illuminate the route.

John Colton's prodigious collection, now the Jayhawkers of '49 Collection in the Huntington Library, is invaluable to historians or writers about early Death Valley history. His collection came to the library from Edward T. Colton in April 1930 and June 1931. The Asa Haynes diary was a gift from G. William Hume in 1957. The materials span from 1849 to 1952; the Huntington Library organized it in 2000. The latest Huntington Library acquisition is what remains of the original copy of Sheldon Young's diary, a gift from his great-great-grandchildren David Duerr and Gretchen Forsyth. Because the Jayhawkers in '49 Collection is closed, Young's diary is located in Manuscripts, call numbers mssHM 75663–75664.

Colton was deeply affected by his experiences in the Nevada-California deserts and his deliverance from that physical and emotional hell. For Colton, the words *Death Valley* took on a far larger scope than the valley's physical size. They became a metaphor for his awful struggle from near death to a long and successful life. He was driven to learn more about the area and their route, and he plumbed the memories of his companions, searching for answers. He spent a huge amount of time, money, and energy to find out and save as much as he could about the Jayhawkers' Death Valley experiences. He initiated the Jayhawker reunions and encouraged the men (and one woman) to attend and, more important, to send letters of their remembrances to be read at the reunions. Due to Colton's dedication, we now have a deeper knowledge of the route and human suffering and compassion the Death Valley '49ers experienced.

Material from the collection should be used with a major caveat: most of the letters are reminiscences written or told many years after the participants' trip through southern Nevada and California. As time went on, memories were jogged, but additional recollections were often influenced by what other '49ers wrote or said. As Dow Stephens wrote in 1887, "I met Tom Shanon & Lewis Manly here in town a few days ago. of course we had to talk over some of the events of our Desert experiences which brings fresh to mind things I had almost forgotten." As the fictional character Sheriff Walt Longmire said, "I think I know things from my past, but it turns out I just think that I know them; my youth is becoming a mythology to me" (C. Johnson 2015, 94).

Scrapbooks 1 and 2 in the collection contain many newspaper clippings (some without sources or dates) written about the various Jayhawker reunions. Many of them are dubious references: They often repeat what was said in other newspapers, or they parrot what was reported about past reunions. Rarely do they contain fresh and valuable material (although the *Beaver Valley Tribune*, February 11, 1898, is an exception). Letters that were reprinted in the newspapers had spelling corrected and punctuation changed, resulting in differences from the originals, such as Brier's "Muskeet Swamps" turning into "muskrat swamps" (Brier 1876) and Colton's death either going "to" or coming "from" his ranch depending on the article you read.[4] These mistakes are then perpetuated in the Death Valley literature. Thus, I use the reunion newspaper articles judiciously.

Colton's handwritten correspondence is difficult to decipher, and his tissue-paper copies (dampened paper pressed on the fresh-ink original) are doubly so. John Groscup wrote Colton (1904), "I received yours of April 7th 1904 A few days Ago but have not ben abel to read all of it yet and have had three other[s] to try to read it but thay failed as well as my self. Please try to wright more Plain in the Futer." Genne Nelson did an expert job of transcribing the old penmanship.

Another problem was the initial reticence of the participants to speak of their travails. Dow Stephens wrote, "These are Scenes I dont like to write about, too horible to contemplate" (1881). John Wells Brier wrote of his mother, Juliet, "It gives her the most excruciating agony of mind to attempt a recital of events connected with our journey. Besides, those events are so mingled and confused in her memory, that her testimony is practically worthless" (1908). Juliet was ninety-four at this time, several years after she wrote coherent letters to the reunions and hosted three of them. But once the floodgates were opened, the relief and comfort found in sharing with others who were comrades in their suffering produced books and articles and meetings where they could share their experiences and tell their stories firsthand.

The next most important references are William Lewis Manly's book *Death Valley in '49*, published in 1894, and his serialized account "From Vermont to California," published monthly from June 1887 through July 1890 in *Pacific Tree and Vine*, which was later called the *Santa Clara Valley: Pacific Tree and Vine*, a horticulture magazine (1887–89).

Lewis Manly was exceptionally well prepared to write a chronicle of the Death Valley portion of his trip to the goldfields.[5] First, his means of making a living by hunting game and evading Indians in the western woods of Wisconsin sharpened his awareness of "place." Then he took the responsibility of scouting a route through the complex Basin and Range Province of Nevada and California for the Bennett and Arcan wagons.[6] Manly (with John Rogers) traversed three times the country from Death Valley into the Santa Clarita Valley in Southern California—down and back to find and deliver supplies to the families in Death Valley and then down again to lead them out.

Manly also kept a diary of his travels. After he returned to Wisconsin in 1853, he was housebound during a severe winter and filled his time writing a three-hundred-page letter for his family in Michigan based on his diary (Manly 1888). The writing process again reinforced his memory of the trip. After the long letter had been read and reread by family and friends, both it and the diary were tragically destroyed in a fire.

Manly again returned to the Panamint Mountains, the western boundary of Death Valley, in winter 1860–61 to rescue Charles Alvord. He spent about four months with Alvord, William M. Stockton, Caesar Twitchell, and his old hunting companion Asahel Bennett, prospecting for "silver mountain," the source of a piece of silver float one of the '49ers found and the source of ore samples Alvord had collected earlier in the season. Manly also traveled by stagecoach from his home in Santa Clara (near San Jose) to Los Angeles and again by train, so he refreshed his memory as he followed part of his route via carriage.

In 1877 Manly was interviewed by a newspaper reporter about his trip, and then years later he was induced by friends to tell the story of his escape from the California desert—from Death Valley. The desert portion of "From Vermont to California," published during 1888, is reprinted and annotated in *Escape from Death Valley*. We think "From Vermont" provides details in a more straightforward way than in *Death Valley in '49*.[7]

In 1890 Manly wrote Henry Clay, who was hosting the 1891 Jayhawker reunion: "I am often asked to put this rough job in book form but I have not yet determined to do so because I am not well satisfied with my own production on this particular subject. I am no man of letters & never before penned an article for publication so you may not expect all to appear as it really should but wiser heads say it is a part of Cal early history & should not be left to die unknown and deserves a place in the Pioneer Records." We heartily salute his resolve and the product of his writing dedication.

Manly's 1894 classic tome, *Death Valley in '49*, was edited by a young woman, Miss Helen E. Harley, who interpreted Manly as best she could. He often "came with scribbled notes on grease spotted paper bags and would sometimes break into uncontrolled weeping during his reminiscences" (L. and J. Johnson 1987, 12). Harley undoubtedly used "From Vermont to California" as a reference, since the Jayhawker portions of the two accounts are virtually the same.

Most of Manly's short articles, republished in *The Jayhawker's Oath* (1949), were written in 1893 or after, about the same time his book manuscript was written, but deterioration in his memory or the need for brevity called for generalization in the *Oath* articles. He was also invited to the Jayhawker reunions and wrote long letters to be read to those who attended.

Jayhawker Lorenzo Dow Stephens wrote numerous letters to the reunions and also penned his reminiscences in a book. Historian Carl I. Wheat wrote, "In 1916 Lorenzo Dow Stephens printed at San Jose his 'Life Sketches of a Jay Hawker of '49,' a sixty-eight page pamphlet which should be read with some caution as Stephens was nearing the age of ninety when it was prepared and his memory was apparently then none too good" (1939a, 115).

The Reverend James Brier provided correct names and locations in his letters to the reunions after he was "induced to go back as far as Death Valley (where you burnt your wagons) as a guide to a Company of Miners" in 1873, and these letters are an important resource (James Brier 1876, 1879, 1898). The *Los Angeles Times* article of a talk he gave in 1886 is also helpful. James Brier, with his son John, planned to publish an account of their trip through the western deserts. The *Los Angeles Times* article said, "He has the Death Valley episode ready for publication in a book of 100 pages, but at present lacks the means to bring it out" (1886). It is possible it became the basis for John's articles in *Out West Magazine* in 1903 and the *Grizzly Bear* in 1911. John sent what he had written (with pictures) to Paul De Laney in 1908 for use in his *Death Valley Magazine* at Colton's suggestion, but De Laney never used it and did not return the material (John Brier 1908). John remembered "the events stamped on the memory of a very small boy" (1908), although he had surely heard the story told many times throughout the intervening years. When John needed to give names to specific places, he was at the mercy of secondhand interpretation and often attached the wrong name to a location.

Juliet Brier's letters give insight into the emotional impact of the trip and provide details of her family's ordeal. Later in life her letters were almost illegible.

After John Colton, the next serious researcher on the Jayhawkers was John G. Ellenbecker, an amateur historian from Kansas who did a very good job with the materials at his command. His *Jayhawkers of Death Valley* was published in 1938 and reprinted by his descendants in 1993 with a five-page supplement of material Ellenbecker wrote sometime before his death in April 1945.

This is an appropriate time to interject a cautionary comment: J. M. Vincent wrote about the pitfalls of historical research, "The inclination to accept everything written, not only as genuine, but true because written, perpetuated the errors as well as the falsehoods of the past. There exists even yet the temptation to accept the words of predecessors without sufficiently weighing their value" (1911, 120). History analyzed or "remembered" after the fact is imperfect, but generally it can be improved by additional data. Much has been discovered and reanalyzed since Ellenbecker's groundbreaking work, so his work must be used after assessing years of additional data.

The hand-drawn map Lewis Manly sent to T. S. Palmer in 1894 is titled "Showing the Trail the Emigrants Travailed [*sic*]: Salt Lake to San Bernardino in 1849." When Manly wrote Palmer about the map, he said, "The sketch I sent you I fixed up before I put my narrative in Book form to show my many friends who made frequent inquires about what we then called the cursed hole (1894 NM59802).[8] LeRoy and I call it the "Palmer Map," one of several maps Manly drew of his 1849 desert route.[9] It is not *to* San Bernardino, although it has "The Spanish Trail Leading from San Bernardino to Salt Lake" along the right side. The trail is *to* today's Santa Clarita Valley, eighty air miles northwest of San Bernardino. The map is not to scale, and the shape and size of the paper forced Manly to draw the north-south mountain ranges west to east and the final leg into the Rancho as south instead of west. He sketched it first in pencil and then in pen on a single page 18 1/8 by 22 3/4 inches. Added to the map in light pencil by an unknown person (probably Palmer) are modern names: "Enterprise Ut.," "Indian Springs Nev.," "Redlands Canyon," and "Goler Wash." We conclude the latter two labels are on the wrong canyons. The Bennett-Arcan party exited Redlands Canyon, not a canyon north of it as indicated by the secondary "Redlands Canyon" annotation. Manly used a group of dots to represent camps and very helpfully labeled several places along the route such as "Found Ice" and "Salt Water Hole" in today's Indian Wells Valley; "Providence Springs," "Dotys Camp," "Trail Forks" (to "Walkers Pass"), "Willow Corral," and "Robinson died." He labeled Death Valley and "Whare Jayhawkers camp/Burnt their wagons" plus two salt lakes. He also drew two salt lakes in Panamint Valley, the valley west of Death Valley.

The decidedly northern turn on the map after the Death Valley '49ers left Mount Misery corresponds with his description of those next days traveling northward, but it totally confuses the issue when one tries to compare it to Bennett's trail used by the Mormon Expedition in 1858 with Bennett as their guide. Manly used numbers on the map that correspond to the following legend that he added (ditto marks removed): "1) Mount Misery here the first death occurred; 2) On Bennetts Arcanes trail; 3) On Jayhawkers Trail; 4) Whare Culverwell died (337 feet below the sea); 5) Whare Bennett, Arcane and their families camped; 6) Whare Isham died (while Manly and Rogers went for help); 7) ["Whare"] Fish died; 8) ["Whare"] Robinson died; 9) Mountain Meadows." If one turns the paper while reading the map, it becomes a very helpful aid for tracing Manly's trail and the Jayhawker trail, but one must balance the advantages with the deficits found in Manly's maps.

After Ellenbecker, other noted historians—Carl I. Wheat, Burr Belden, Charles Lummis, Dale Morgan, and LeRoy Hafen and Ann Hafen—have worked on the Jayhawker history. All have contributed to it as they gained access to more material.

Genne Nelson, who has lived and worked in Death Valley for many years, became intrigued with the '49er story when she was curator for the Borax Museum in the heart of Death Valley (which was started before the Park Service had a visitors center). Genne meticulously transcribed Jayhawker material in the first three boxes and the first two scrapbooks of the Huntington Library's Jayhawkers of '49 Collection, and she kindly furnished me copies of her hard work to help prepare this book. As she transcribed hundreds of items, she became an expert at deciphering difficult handwriting, and as a scientist (a geologist), she was exacting in her attention to detail and annotation of sources. LeRoy and I have verified Genne's excellent work with microfilm and photocopies of original documents in our possession. We have also visited the Huntington and the Bancroft libraries over the past forty years to read the original letters and diaries.

Late in life John Colton had planned to write a history of the Jayhawkers (E. Mecum 1908). In 1908 he wrote Dow Stephens and John Brier to encourage them to write something for Colton to use. Stephens replied:

> I send enclosed a few more sheets that I have blocked out from memory and intended to revise it and get it in little better shape but find it too much work so just send it as it is. If you can make any use of it do so. you will see I have got us far as Poverty Point [Mount Misery] where they all turned back so if the story is going to be published I will commence where I left off and go on through but I have not heard a word more about it. pleas Inform me If there Is any thing being done. (1908)

Apparently, the publication Colton had in mind was not a book but a series of articles to appear in the *Death Valley Magazine*, published in the boomtown of Rhyolite, Nevada, east of Death Valley. Ironically, the editor had placed the following ad in the *Editor* in January 1908: "Paul De Laney, advises us that he is in the market for a small number of desert stories not exceeding 1,500 words. These must be confined to the desert and be thrilling or pathetic stories of truth." Colton's story was both thrilling and pathetic, although not completely true, and it ran far more than fifteen hundred words. The first issue of "Jayhawkers of '49" carried the byline "written by Col. John B. Colton, edited by Paul De Laney." The next issue, June–July, continued with "Lost in Death Valley, . . . written by Paul De Laney from notes of Col. John B. Colton." It includes a picture of young John Colton at age seventeen when, John later claimed, "he organized the first detachment of the Jayhawkers." This statement stretched the truth considerably, as did the declaration in the August issue that a teenage emigrant whippersnapper (Colton) could influence a confrontation between a headstrong and tipsy Jim Bridger and the powerful Brigham Young in Salt Lake City. "I mounted my horse," claimed the now aged Colton, "and joined Bridger, warning him not to cause trouble." The September–October issue covered the Jayhawkers' stay in Salt Lake City and their travel on the Southern Route to Mount Misery. This issue ended with Captain Hunt saying, "Boys, I'm sorry, but you are all going straight to hell." That was literally "The End"—the last issue of the *Death Valley Magazine*! One can only speculate whether De Laney just quit publishing or if Colton no longer provided information for him to work with. Two years later, De Laney appeared in Denver and lived until 1946 (*Spokane (WA) Daily Chronicle* 1946). I find no indication Colton tried to find another publisher for the rest of his story (if there was one). Fourteen years before, he had written T. S. Palmer, "It is a long story, too long to write—I have collected the records, letters & Everything...but life is too short to write it" (March 21, 1894).

Notes for Appendix F

1. Lorton was married on July 16, 1862 (New York City Marriage Notices, 1835–80; findmycast. com). The children were Edith (August 17, 1863), Claude (May 25, 1868), and Belle Louise (July 25, 1872). Mary already had a young son, William Briggs (April 8, 1855).

2. Colton had this typescript made in 1897, although his secretary undoubtedly typed it. Margaret Long published it in her two editions of *The Shadow of the Arrow* (1941 and 1950).

3. A signature is a unit of a book that was originally a large sheet of paper with several smaller pages of printing on it, front and back, then folded, stitched together, and cut into book-size pages.

4. "To his farm" (*Rocky Mountain News*, October 30, 1919); "home *from* his ranch" (*Daily Illinois State Register*, October 27, 1919).

5. Manly used his middle name as his given name, as did his family and friends. We honor his choice. Examples in *Death Valley in '49*, 22, 23, 125, 151, 195, 221.

6. Various '49ers spelled "Arcan" as "Arcain," "Arcaine," or "Arcane," implying the pronunciation was with the long *a*. We use the "Arcan" spelling found on the family tombstones.

7. This account is bylined "W. L. Manley," as was the name for his farm south of San Jose on the 1876 Thompson and West map. It is on the last page of *Death Valley in '49*, for his desired obituary. Since the two names apply to the same man, I have standardized the spelling as "Manly" to reduce confusion.

8. Manly wrote Palmer on December 14, 1894, "I drew the map in 1890 while my Book was being printed" (HM59803). Thus, I have used the 1890 date since it fits with Manly's previous statement that the map was "fixed up before I put my narrative in Book form," even though Manly continues that it was "while my Book was being printed," which would put the date as 1894.

9. Manly Map, 1890, to T. S. Palmer. The map is included in Carl I. Wheat's monumental *Mapping the Transmississippi West* (3:106, map 628).

NOTES

Notes for Introduction

1. Colton said at another time the name "was coined on the Platte river, not far west of the Missouri river, in 1849, long before the word 'Kansas' was known.... Hawks, as they sail up in the air reconnoitering for mice and other small prey, look and act as though they were the whole thing. Then the audience of jays and other small but jealous and vicious birds sail in and jab him.... Perhaps this is what happens among fellows on the trail...out of pure devilment" (Fox 1904, 17n).

2. Great Salt Lake City was the first name given to the Mormons' new Zion, a settlement nestled on the fertile land between the Wasatch Range and the Great Salt Lake (O. Pratt et al. [1947], 80). It was later shortened to Salt Lake City, which I use throughout.

3. A mess is a group of three to eight men who eat and travel together, usually with a single wagon in a wagon train.

4. See Appendix B for a list of names. The number of Bug Smashers comes from Evans 1945, 254, 259.

5. An article in the *Kanesville (IA) Frontier Guardian*, May 30, 1849, included the roster, constitution, and bylaws of the Knox County Illinois Company when it left Kanesville on May 23 after the addition of new members that doubled the number of the original company (see Appendix A). William Lorton recorded that the company was formed May 12 more than a hundred miles south of Kanesville and that the roster of the original company was signed on May 16, on their way from St. Joseph, Missouri, to Council Bluffs, Iowa.

6. Ed McGowan, Dow Montgomery, and Robert Price were also invited to reunions and are mentioned in Jayhawker letters, so they were probably Platte Valley Jayhawkers. A note Colton wrote on McGowan's postcard in 1872 gives some information about McGowan.

7. A William Robinson from Maquon, Illinois, about ten miles southeast of Knoxville, is mentioned as traveling with the Jayhawkers (Asa Haynes Manuscript Diary, 1849–1850, JA1051; Clay, obituary, December 14, 1897).

8. There was also a William Robinson in the Colony Guard from New York, from which thirteen men took the Southern Route to Los Angeles (B. Madsen 1982, 63, 68–69; 1983, 25–26). Of those thirteen, Edward Coker, Samuel S. Abbott, and possibly others became Death Valley '49ers, although not Jayhawkers (Nelson 2002, 270). Coker traveled with the Bug Smashers, and Abbott became one of Asahel Bennett's teamsters, probably from Mount Misery into Death Valley. He left the Bennett family in Death Valley to become part of the northern contingent. Coker gave his very garbled account to Manly sometime before May 1892 (Manly 1892; Coker in Manly 1894, 147, 373–76). Although one cannot rule out the New York Robinson as the Death Valley Jayhawker, I lean toward the Maquon, Knox County, Robinson as the right one.

Notes for Chapter 1

1. Gold was not found on the banks of the Sacramento, but it was found along its tributaries. Until gold camps were named, Sacramento was the goal.

2. This letter was presumably from the goldfields, dated November 24, 1848.

3. A tea box with interior dimensions of about 3 x 4 x 4 inches would have held the 230.769 troy ounces of gold dust and nuggets sent to Washington, DC. A solid gold bar weighing 230.769 troy ounces would fit a space 3 x 3 x 2.52 inches and weigh 15.824 avoirdupois pounds.

4. Colton's mess left Galesburg on April 1, but they spent a rainy five days at the Clay farm three miles west of town. A "mess" is a group of men who travel and eat together.

5. Galesburg wrested the honor of county seat from Knoxville in 1873 when Galesburg became a major railroad hub.

6. Asa Haynes dictated reminiscences to his daughter Nancy Haynes Wiley at an unknown date. Years later, on June 28, 1937, Nancy Cramer (Nancy's granddaughter) made a typescript of Wiley's handwritten transcription.

7. This considerably garbled narrative of Asa Haynes's memoir was taken down as notes by his daughter Nancy J. Haynes Wiley and rewritten by Ruth Cole, another of Nancy's grand-daughters.

8. William Hall Kellogg's birth is variously recorded as 1827, 1828, and 1830.

9. Shockcon is Shokokon (Shokoquon), Illinois, an unincorporated village on the east bank of the Mississippi River. "It is only a river landing now [1911], but in the early days it was a thriving settlement, doing a good business in shipping and merchandising" (Gordon 1911, 633).

10. N. D. Morse is listed in the Knox County Company roster. A man named John Morse is often listed in the Jayhawker reunion lists. At this time I assume "N. D." and "John" are the same person. "Oren Clark" was listed in the Knox County Company roster, and a Charles Clark was a Death Valley Jayhawker. Oren and Charles could be the same person. James Ellery Hale, however, took the Northern Route with Luzerne Bartholomew and the Kelloggs (Hale 1850).

11. Edward and his older brother Luzerne Bartholomew were in the train after April 10 and signed the constitution roster. Neither the father (Noyce) nor one of Edward's brothers (also named Noyce) was in the train after April 10.

12. One of the men sent ahead may have been Robert Taylor, whose name is near the end of the company roster that was completed in Kanesville. E. Norman Taylor, also on the roster, traveled with his wife and young son. On April 6, 1848, the hamlet in Millers Hollow was given the name Kanesville to honor Thomas L. Kane, a non-Mormon emissary of President James Polk. Kane helped negotiate federal permission for the Mormons to use Indian land along the west (right) bank of the Missouri for their winter encampment of 1846–47. The general area around Kanesville was called Council Bluffs, although the council site used by Lewis, Clark, and the Otoe and Missouria Indians was twenty miles upriver at today's Fort Atkinson State Historical Park, Nebraska.

13. Kanesville's name was officially changed to Council Bluffs in 1853, but the area had been referred to as Council Bluffs for many years. The boat with their baggage got stuck below St. Joseph and never did get up the Missouri to Kanesville.

14. The rear wheels were prevented from rotating by chaining the wheels to the axle or by inserting a felled tree or heavy branch through the spokes. Occasionally, a rope was attached to the rear axle, and men slowly lowered the wagon while another man controlled the lead oxen (Wilkins 1968, sketch 8, "Crossing a Creek").

15. Mormons established Mount Pisgah, Iowa, in May 1846 as a way station where Mormons fleeing from eastern states and Nauvoo, Illinois, could rest and resupply during their trek to their yet-to-be-located Zion (Gentry 1981; Hartley and Anderson 2006, 86–90). Sheldon Young, who later became a Death Valley Jayhawker, used the same route as the Knox County Company through Iowa, but he was about two weeks ahead of them. He stayed on the south side of the Platte River from the Missouri to Fort Laramie.

16. The headwater of the little Platte River is in Iowa and flows southward. The river's official name is "Platte River," which is at variance to the "Little River Platte" name shown on William Clark's map (Map 1814). Lorton's "quite a settlement" of three log cabins is an example of his tongue-in-cheek sense of humor.

17. Robert Price did not go through Death Valley, but he was a Platte Valley Jayhawker and remained a good friend of Colton throughout their lives. Alonzo Clay did become a Death Valley Jayhawker.

18. Ray was apparently with the Findley train. The Knox Company was stopped because the ferry rope had been cut when it became entangled with the paddle wheel on a riverboat. It took a couple days to get another rope secured across the swollen river (Lorton April 17).

19. The list of members printed with the constitution and bylaws was not complete. One hundred six men are listed, but at least one woman and a child are not on the list, and several men are not listed, including Luther A. Richards and William Robinson, both of whom became Death Valley Jayhawkers. Rasmussen (1994, 1:64–65) does not include William Lorton, Joseph Ellis, Jesse Workman, Luther Richards, or William Robinson. I suspect Robinson was in another wagon train.

20. According to Meredith Runner (1995), Luzerne's '49er experiences were "published by his biographer, Charles Reed, in *Sports Afield* magazine about 1888." The brothers arrived in Salt Lake City in the Knox County Company—their names are listed together in "Emigrant Rosters 1849" of the LDS Church for August 8, 1849 (copy in Cannon 1999, 146; see also L. Bartholomew 1905).

21. "Indian country" is described in the 1834 *Statutes at Large*, 4 Stat. 729, as that country west of the Mississippi River "and not within the states of Missouri and Louisiana, or the territory of Arkansas."

Notes for Chapter 2

1. There were three men with the initials J. B. listed in the Knox County Company roster: J. B. Anderson, J. B. V. Wallace from Galesburg, and J. B. Surmeier from Adams County, but these men are not on any Jayhawker list. Possibly Mecum miswrote "J. L." for J. Leander Woolsey or "J. G." (John Groscup), "E. B." (Edward Bartholomew), or "B. B." (Bruin Byrum), who were all Jayhawkers.

2. The term *boys* can be interchanged for *guys*; it had nothing to do with race or age.

3. There is no record of Isham being in the Knox County Company. There is a reference to Ed Doty possibly having a violin. "Mr. Doty (Ed's son) told of their taking his father's violin,—which was too much of a burden to carry further [in the desert]—and used it to fry some meat and have a 'Musical Steak.'" John Brier said Nat Ward played violin, and Lorton also played violin in the Knox County Company (E. Mecum 1931; Brier 1903, 331).

4. A rod equals 16.5 feet; thus, the Platte was about 550 yards wide (about a third of a mile) here.

5. Scouts had warned the drivers they would be crossing a marshy area, so they carried brush to make a firmer roadbed. They also carried brush for campfires because firewood was scarce and there might be few dry buffalo chips (dung) ahead to burn.

6. This may be Thomas Gordon from Mercer County, Illinois. Boyd may have been a nickname (see membership roster in Appendix A; Rasmussen 1994, 65).

7. Ash Hollow is on the south side of the North Platte River, two miles southeast of today's Lewellen. The grave was on the north side of the river, the same side as the Jayhawkers' train.

8. The Ancient Bluff Ruins along the north side of the North Platte River are just east of Broadwater (about twenty-five miles west of Oshkosh, Nebraska). The picturesque formations along the North Platte are "made of stacked layers of sandstone, siltstone, mudstone, and volcanic ash. Differential erosion of these layers—some soft, some hard—gives each monolith a distinct shape" (Meldahl 2007, 59).

9. More astonishing than the rock itself is this: If you mentally project a horizontal earthen plain from the top of Chimney Rock in every direction, all of the empty space below this projected surface represents silt, sand, rocks, and volcanic ash that has vanished by wind and water erosion—some of it helping to form the Mississippi Delta.

10. Dissension occurred just as Findley had predicted and Lorton had reported from the meeting he attended the previous February.

11. Frémont camped on Deer Creek July 26, 1842, and favorably described where it empties into

the North Platte: "Here was an abundance of rich grass," and the stream was "well timbered with cottonwood of an uncommon size" (1970, 238).

12. Craig Bromley, archaeologist for the Bureau of Land Management at Lander, Wyoming, used detailed topographic maps and supplied these data: "The very highest point on the east side of the Gate is 361 feet above the river. This point is about 65 feet back from the edge. The highest point on the east side of the Gate where you can almost look down onto the river is 353 feet above the river. This point is about 15 feet back from the edge." Personal communication with LeRoy Johnson, December 9, 2013.

13. Lorton was gradually recovering from his stampede injury. He now walked with a cane.

14. Because of a broken axle from a stampede on the July 18, John Colton missed this one on the 19th but mentions it in his letter started July 17 and finished the 24th.

15. Henry J. Ward was in Lorton's mess. He had worked for Lorton's cousin Silas Olmsted at his mill near Monmouth, Illinois.

16. Lorton wrote on July 14, "Waggons were smashed, men knocked down. Canton man broke 2 ribs. old man [E. Norman] Taylor had his wagon to pass over his boddy, but did not severely injure him."

17. Colton's first curiosities are true, but his following comments are facetious. Guaiac resin is from *Lignum vitae*, a southern species not native in Wyoming, but Colton would not know that.

18. This is not quite accurate. Runoff between the Rocky Mountains and the Wasatch Range to the west drains into the Pacific. West of the Wasatch, in the Great Basin, water drains to several low points, such as the Great Salt Lake, Sevier Lake, and the sink of the Humboldt (Mary's) River.

19. This was the height of hyperbole. As Lorton said July 25, "We come to the South Pass & dont hardly know it, the assent being so gradual & the ground being so rolling.... We expected to see a deep pass between high rocks & only capable of admitting a single wagon, [but] several roads can be made thro' here."

20. Emigrants mounded earth or piled stones on a grave to thwart wolves and coyotes from digging up the body.

21. A tepee was the classic cone-shaped dwelling of the Plains Indians.

22. William Clayton and Thomas Bullock are likely the first diarists to record the existence of the Cache Cave, both on July 12, 1847 (Clayton 1921, 291–92; Bullock 1997, 225). For more about the cave, see Kimball 1988, 96–97, map 7; and Berrett and Anderson 2007, 246–54.

23. The Mormon Church coinage system was supervised personally by Brigham Young. In 1851 the U.S. Mint assayed the Mormon $10.00 coin and claimed it had only $8.52 worth of gold in it. Soon bankers and merchants discounted the Mormon coins 20 to 25 percent. Today, a Mormon coin rated rare fine and extrafine is valued up to $90,000.

24. "The Mormons wanted a road made through to the city *de los Angelos*, which they looked upon as the commercial Metropolis of their state. It is included within the limits" of the proposed state of Deseret, J. H. Purdy wrote after leaving Salt Lake in September 1849.

25. During the Mormons' initial flight from Winter Quarters to the Salt Lake basin, they perfected a wooden roadometer (odometer) that recorded revolutions of a rear wagon wheel. They knew the circumference of the tire in feet (C), and the roadometer recorded the number of revolutions it made in a day (R). At the end of the day, they could calculate miles traveled (R x C ÷ 5,280 feet per mile) (see Clayton 1921, 136, 143, 152).

Notes for Chapter 3

1. Captain Hunt and fellow Mormon Battalion soldiers purchased two small cannons from John Sutter after the soldiers were discharged from the U.S. Army at Los Angeles in July 1847 (Ricketts 1996, 169, 203, 330n28).

2. The number 430 miles on the Green River is the distance from the Oregon-California Trail crossing in Wyoming (the Mormon Lombard ferry at the confluence with the Big Sandy) to the Old Spanish Trail crossing above today's town of Green River, Utah (Kane 2008, iv, 3, 94, 203, 208).

3. The Mormon wagon had an odometer. The distance is the same from Salt Lake City to today's Adamsville by current highways, according to Google Maps.

4. Another group of about twenty packers, mostly Mormons under the leadership of James Flake, traveled in tandem with the Smith packers.

5. There were additional inferences about Indians using horses for food rather than transportation. Young, December 20, 22, 1849, January 6, 1850; Manly 1894, 165.

6. A "note" on the 1846 map says it "was one of Mitchell's most popular and important pocket maps." Map 1846 (see in Wheat 1959, map 520, facing p. 29, and in Woodward 1949, facing p. 168); Map 1849, Mitchell.

7. On October 25, Welch notes "the train was soon leaving the 'Valley of Errors' [Escalante Desert] taking the cañon through which we had come [just east of today's Minersville]. The oxen are quite weak and need rest with grass & water" (Ressler 1964, 268).

8. Jeremiah Fish said in a letter dated March 1, 1850, from Mission San Gabriel, "The 7th [division] was the first of our train that got into this valley; they brought in but five of their waggons and about one fourth of their cattle.... The 1st Division [Buckskins] was the next in with eight teams, and a little more than one-third of their cattle.... Our division next: We arrived with 15 teams and 81 head out of 195; the rest came scattering along and out of provisions."

9. Today the hills are covered with piñon, but LeRoy Johnson bored the trees to count their growth rings and found that with few exceptions the forest is younger than 150 years.

Notes for Chapter 4

1. For an in-depth look at past research of the Mount Misery area, see L. Johnson 2005, 43–74; and 2018, 10–43.

2. All these men except Lorton became Death Valley '49ers.

3. See also Hoshide 1996, 210, 212; Elliott 1883, 176; Kelly 1939, 41; and Evans 1945, 254, 259. Osborn's inscription is in the first westward gulch out of Headwaters Wash, a half mile below Tunnel Springs. LeRoy christened it "Osborn Gulch."

4. Tom Sutak made arrangements to have John Grebenkemper and his cadaver-sniffing dog, Kayle, investigate a rock pile on the side of Roundup Flat that looks like a grave site. "Kayle DID NOT alert on the rock pile or anywhere else on that ridge. John says that she showed 'no interest' anywhere up there. In John's opinion, it's not likely that there is a detectable burial there" (Sutak, March 21, 2016, personal communication). We now assume Naylor was buried in softer soil south of Roundup Flat, where cattle and wild turkeys trod among the sagebrush.

5. Lorton planned to rendezvous with five men from New York City, his hometown: "Charles Burrel, Wm. Sands, Wm. Sherman, J. Hendle, J. Bucklin" (January 20, 1850).

6. Lorton fortunately recorded the names of the Pinney-Savage party for posterity: Chas. McDermot, Mr. Savage, Jno. Adams, G. Wiley Webster, T. Ware, J. Ware, Mr. Baker, Mr. Semore, Mr. Allen, and Mr. Moore. Ibid. He mentions Pinney in his journal, Nov. [20] (corrected date), but not Mr. Moore.

7. See Manly 1894, 355; Lingenfelter 1986, 39–40; and Stephens 1916, 21. Seeley, a member of the Pomeroy wagon train, recalled they "picked up nine men that had at one time formed part of the Company that suffered and perished in Death Valley" (in Hafen and Hafen 1954, 296). But other journals make clear these men were some of O. K. Smith's packers who were heading back to Salt Lake City and were able to join a train and continue to California (Hamelin and Thurber in Hafen and Hafen 1961, 109, 117–18).

8. Manly sketched a "map of my own make to show you [John Colton] according to my memory the different trails we made" on which Manly labeled "Mt. Misery" and added, "Where many turned back only 26 wagons took the cut-off" (February 17, 1890; map, ca. 1889). The map was redrawn to fit one page (Long 1941, 161; 1950, 170; see also Manly December 1887; and 1894, 113).

Notes for Chapter 5

1. Jacob, John, Henry, and Abraham Earhart, from Johnson County, Iowa, are listed by Stover as traveling in the Sacramento Mining Company from Iowa City (Stover 1937, 166). Jacob and his son Abraham died of smallpox in Sacramento just two days apart, November 27 and 29, 1852, and are buried in the Sacramento City Cemetery. Jacob's brother John was in Sacramento for the California census taken two weeks later on December 12, 1852. There is no record of Henry having been in California. California State Census 1852; Burial Index—Sacramento Historic City Cemetery 1852. See also Manly 1894, 113, 133, 153.

2. After Colton started the Jayhawker reunions in 1872, he created his Death Valley Jayhawkers list by adding to the Platte Valley Jayhawkers several other people who traveled with or near the Jayhawkers through the deserts—the Brier family, Sheldon Young's mess, Isham, Father Fish, Dow Stephens, the unnamed Frenchman, and a man named Gould, apparently from Oskaloosa, Iowa.

3. Juliet died on May 26, 1913 (San Joaquin County Cemetery Records 1893–1967; San Joaquin County Death Certificates Index, 252; obituary, May 27, 1913).

4. "I, Aaron Woods, an ordained minister of the church and authorized to solemnize marriage in the State of Michigan, certify that I joined as husband and wife, James Brier of Covington, Indiana, aged twenty-five years and Juliet Wells of Mottville, Michigan, aged twenty-five years . . . on the twenty-third day of September, 1839, at the house of John Wells in Mottville Township" (James Brier 1838).

5. Several men listed in the Badger Train became Death Valley '49ers: William Abbott, A. C. St. John, Asahel Bennett (and family), and possibly Anderson Benson (Rasmussen 1994, 1:67). See also Appendix B. The R. Moody and M. Skinner families, who became friends of the Bennett and Arcan families (Death Valley '49ers), turned back to the Old Spanish Trail at Mount Misery. They all started from the Missouri River in the Badger Company. Abbott and St. John were teamsters for Bennett, but they left the family in Death Valley to follow the Jayhawkers.

6. Carl I. Wheat uses an incorrect date for James Brier's birth (1939a, 108).

7. Columbus was named for his father's brother, not for the explorer (MacBrair-Koller 1988). Kirke White's state of birth is from California 1852 and U.S. 1860 Censuses. It is recorded as Ohio on several genealogy websites.

8. James Brier's letters written to the Jayhawker reunions in 1876, '79, '93, '96, and 1898 are reproduced in L. and J. Johnson 1987, 176–87.

9. His name may have been Coverly, according to Ellenbecker, or it may have been Jerome Bonaparte Aldrich, who later lived in San Bernardino, according to T. S. Palmer (Ellenbecker 1993, 53; Palmer in Wheat 1939a, 105). Also in the Nusbaumer mess was a Negro man named Smith who turned north with Fish, Isham, and Gould. He may have joined the Bug Smashers in Panamint Valley. He is not mentioned by the Jayhawkers.

10. Lorenzo Dow Stephens was sent an invitation for the 1879 reunion (JA841) and wrote numerous letters to them over the years. The 1901 and 1918 reunions were held at his home in San Jose, California. (Colton 1901; C. Mecum 1901. See also Appendix C).

11. Colton later learned Young had not died in California but had returned to the states and died forty-two years later in Moberly, Missouri, on August 18, 1892. Stephens thought "the Man from Oskaloosa name not given I think is the Old fellow that had the Mule & Oxen and packed them all the [way] up the Coast until he got to the Mines. his name was Gould and I think when he firs[t] joined us he wore a plug hat" (Stephens 1884). Wheat credits Stephens with saying Gould died at Pen Yan, New York, near the close of the 1850s (1939a, 105).

12. Gritzner was from Joliet, Illinois. T. S. Palmer had been a member of the Death Valley Expedition in 1891 and much later edited *Place Names of the Death Valley Region in California and Nevada* (1980).

13. In the December 1887 issue of "From Vermont to California," Manly says they drove north

three days, but in his book (1894, 124) he said they camped beside a lake after about three days. This does not fit anything in Sheldon Young's diary: more data difficult to reconcile.

14. John Brier (1903, 331) names Nat Ward as the fiddler, but Isham may also have had a fiddle: "old Mr Isham, the fiddler." By "single men" I mean men who did not have family members with them.

15. Manly consistently mentions a map by Williams, presumably mountain man Old Bill Williams, who had been killed by Ute Indians on March 14, 1849, months before the emigrants arrived at Salt Lake City. Thrice Manly clearly confuses the "Williams" map with the Barney Ward map that packer O. K. Smith carried (1894, 109, 110, 325). My suspicion is Manly simply mixed up the names of Williams and Ward. Manly's "pillars of sandstone" are erosional features of the Panaca Formation, sediments deposited in a middle Pliocene lake about three million years ago (Tschanz and Pampeyan 1970, 76–80).

16. According to Young's diary, it took thirteen days to reach "Misery Lake," and he mentions no dry lake before that. A "sightly elevation" is where a good sighting can be taken.

17. Two books on Nevada place-names do not give a source for naming the Seaman Range. It was probably a person's name, but it is fun to think it may have been named for Sheldon Young, the first sailor to venture into the area.

18. This information is also found in James Brier 1886; John Brier 1903; Juliet Brier 1913; and Belden 1954, 29–35. Juliet continues: "Three days later we reached a branch of Forty-mile Canyon, where a foot of snow fell upon us [at Cane Spring]," which implies her "miserable lake" is Groom or Papoose Lake. (The upper reaches of Cane Spring Wash funnel a person into Jackass Flats acting like a branch of Fortymile Canyon.) One must be very careful with recollections after fifty years that may have only a rough chronology (if any at all). Both John and Juliet likely used James's "Death Valley: Its Ghastly Story as Told by an Aged Survivor" (1886) as a main reference for their recollections.

19. Young's diary does not account for "2 or 3 days to recruit," at a warm "stream and plenty of grass," but on November 19 he records, "Came to plenty of grass & water...had the best grass since we left the Platt." This would be a logical time to recruit. There is no guarantee Young's dates are correct until Christmas, and some groups may have lagged behind while others forged ahead a day or two.

20. Both the Panamints and the highest peaks in the Sierra Nevada are visible from Highway 95 in upper Amargosa Valley. We have not been on the restricted parts of the Nellis Air Force Range to confirm their visibility from Emigrant Valley or surrounding high points.

21. Towne Pass, over the Panamint Range, is not visible, but the depression in which the pass is situated (between Pinto Peak and the Cottonwood Mountains) is visible from a great distance. Here Manly may be referring to all the snowy parts of the Panamint Range, not just Telescope Peak.

22. When trying to adhere to Young's mileages and directions, keep in mind he had no roadometer; thus, he recorded no fractions of miles. Over many segments of trail, small inaccuracies can add up to discrepancies of a few miles. The term *ballpark* takes on significance in analyzing the route. I used National Geographic's "TOPO!" mapping software to follow the route and for my choice of place-names.

23. From high points near this section of the trail, such as Tippipah Point, they could see the snowy Panamint Range (and probably the Sierra Nevada in the far distance).

24. The section of Manly's book titled "Story of the Jayhawkers" was compiled by Manly's young editor from "notes and interviews" he collected primarily from three Jayhawkers, Tom Shannon, Ed Doty, and Dow Stephens. She also relied heavily on Manly's "From Vermont to California." Thus, the above information is at least secondhand.

25. Young's "greasewood" is today's creosote bush (*Larrea tridentata*), and his "blue Ash" is now velvet ash (*Fraxinus velutina*).

26. The "2" is very clear. The next number (if it is a number) is scribbled over and impossible to de-

cipher. This camp is extrapolated by counting "6m" back from Young's next camp at Travertine Springs. When they found grass and water, they recruited.

27. Technically, Furnace Creek Wash heads in Greenwater Valley and joins the wash from the pass at the road junction to the Billie Mine and Greenwater Valley. At Navel Spring the '49ers were in the Furnace Creek Wash drainage.

28. Juliet repeats this almost verbatim in her second story reprinted in 1913.

29. Manly also wrote John Colton about the same incident in July 1894. W. S. Barton, a prospector, "found the camp whare Jim Martin left his wagons [Cane Spring]. here the stones that the camp kettle was set upon in 49 still remain in place near the ash pile."

Notes for Chapter 6

1. The Timbisha Shoshone Indians became officially recognized as a Native American tribe in 1983.

2. Lingenfelter (1986) defined Death Valley in a broader sense, as the whole watershed for the Amargosa River. His emphasis was mining history, not '49er trails.

3. Lingenfelter says, "Tradition in the Bennett family gives [Asahel] Bennett the honor" of naming the valley and that it was named in 1860 when Bennett and Manly went back to prospect and find Alvord (ibid., 8, 69–71). Both men were in the right place at the right time to be contenders for the honor. The others were not.

4. Manly stayed with the Halls (Genny's grandparents) near Mineral Point, Wisconsin, in 1853, when he returned from California and where he wrote the three-hundred-page letter about his trip to California for his family in Michigan. Manly also loaned Mr. Hall $600 to "cross the plains to California" (Manly 1894, 437). LeRoy Johnson suggested Genny donate her pristine first edition of Manly's book to the Huntington Library, which she did.

5. LeRoy always carries an accurate Fahrenheit thermometer when he hikes the desert. That day we waited in the shade of a bluff until the sun went down to hike out from McLean Spring.

Notes for Chapter 7

1. The children: Brier family—three; Arcan—one; Bennett—three; Wade—four.

2. The moon gave about 80 percent illumination this night. Young's three plus four plus one miles equals his cumulative eight miles traveled.

3. The distance from lower Salt Creek around the bluffs to *Tugumu* is less than three miles.

4. Van Dorn's articles are not bylined, and the author is identified by deduction.

5. Spears's book abounds with incorrect material about the '49ers.

6. Note to researchers: There are two JA646s in the Jayhawkers of '49 Collection at the Huntington Library: this one (Manly to Bro[ther] Jayhawkers, January 10, 1893), and an undated one to Colton with the envelope stamped "JAN 28 · 93" (in box 2, folder 45). The undated one to Colton I designate JA646[b]. It refers to Manly sending an article he wrote that is cataloged as JA650.

7. Nusbaumer mentions "Hadapp, Calverwell from Washington City, Fisch from Indiana, Isham from Michigan and a colored man named Smith from Missouri—left Salt Lake City" on October 7, 1849 (in Koenig 1984, 33). Gould may have joined them at a later time. Smith traveled in Nusbaumer's mess from Salt Lake City to Travertine Springs, where he left with some of his messmates to join the northern contingent. Whether he continued from Panamint Valley with the Bug Smashers or turned south with Fish and Isham is not known. He is not mentioned by the Jayhawkers. Brier said, "Some of the Mississippi boys, with little West, & Tom & Joe the darkies, left us & steered for a cannon on the west side of the Panamint valley... now called Darwin's" (1879), and he does not mention Smith traveling with Fish and Isham through the Slate Range. Another mystery.

8. Parentheses are in the original. On December 25 Nusbaumer says there was just one yoke of oxen, but the next day he mentions four oxen. On December 28, Nusbaumer, Hadapp, and

Culverwell remained at Travertine Springs, while "our other fellow travelers left us." Fish, Gould, Isham, and Smith turned north to follow the Jayhawkers. Manly says the oxen were left with Nusbaumer, Hadapp, and Culverwell, although the flesh of some was probably dried for backpacking (L. and J. Johnson 1987, 58). For an overview of what happened to Nusbaumer after Death Valley, see ibid., 157–58.

9. Parentheses are in the original. According to Colton, these were the Georgians (April 12, 1894, and November 20, 1895).

10. Dr. George said his prospecting party named the spring "Marble Spring" in 1861 several months before Dr. Owen arrived (1875).

11. "Left our wagons & packed our cattle Thence West 15 mile Thence SW. 24 mile to Snow Canyon" (Haynes Diary).

12 "Juliet Spring" has been informally called Palm Tree Spring for years, but the name is no longer descriptive since the Park Service removed the nonnative palm that consumed most of the potable water needed by local critters. For some reason, this spring is not on USGS quadrangle maps.

13. The Briers had been traveling with or near the Bug Smashers as far as Death Valley and were still with Townes's mess in Panamint Valley after the other Bug Smashers headed west in the northern part of Panamint Valley.

14. Early maps by Van Dorn (1861), Farley (1861), and Wheeler (1877) spelled it "Town's Pass." Van Dorn's 1961 pencil sketches (two sheets) were the first to show Tulare and Death Valleys (L. and J. Johnson 1988, 147-160).

15. Whether those who moved on were all of the Bug Smashers or whether Townes's mess was traveling behind with the Briers is not clear.

16. According to Sheldon Young's diary, he stayed two nights in the Panamints, but he might have been behind Colton's mess. Colton may have been lumping the Mississippi boys in with the Georgians.

17. Pratt mentions Towne as captain of a division in the San Joaquin Company (1990, 381). Nusbaumer also called Townes "Captain" after Mount Misery. "It was decided that ten wagons should drive on together, and to take charge of these we elected a captain named Town." Then they turned toward the west "after chopping our way through a forest the entire day" (Nusbaumer 1967, 34).

18. Most of the northern contingent descended the unnamed canyon that debouches into Panamint Valley just north of today's Highway 190. We, too, have hiked down this pleasant desert canyon more than once. Some of the Bug Smashers climbed to a high point above the pass and met the others in Panamint Valley.

19. Some of the items in the trunk were a pair of worn-out children's shoes, two ceramic bowls with lids, a shawl, a doll, coins, a hymnal, and a letter to "Edwin" mentioning a girl named Lydia and signed William. All the items appeared old but were in astonishingly clean shape, and the trunk lid was glued to the trunk with Elmer's White Glue.

20. Jayhawker Canyon may have been named based on Rude's inscription (Palmer 1980; 2006, 39). Palmer gives incorrect information, but *Dictionary of California Land Names* (1946, 137) is accurate.

21. Haynes's diary refers to "Snow Canyon." Towne Pass, in the four-thousand-foot vicinity, is more canyon-like than the broad top of the pass.

22. I assume "'Old Francis,' the French Canadian" was one person, not Old Francis *and* a French Canadian. He might be the man who carried water up to Mount Misery to sell.

23. The twenty-two men Evans recorded arriving at Aqua Fria is close to the twenty-one number given by Coker, who was one of them. Coker also mentions "friendly Indians" (Manly 1894, 373).

24. In his 1898 letter Brier says the Townes mess departed at "Mesquit swamp," but his detailed 1896 letter is more accurate with the departure at Post Office Spring (L. and J. Johnson

1987, 182, 185). Crumpton and Mastin are not listed in the Pontotoc train from Mississippi (Rasmussen 1994, 8).

25. There was water on the playa that is crossed by today's Highway 190.

26. Young's mileage seems to be the total traveled to date (January 4 and 5) after entering Panamint Valley.

27. This was James Brier's "Indian Springs & Wickeups" (1876), John Brier's "bones of horses" camp (1911a, 4), and Carl I. Wheat's "Horse Bones Camp" (1939d, foldout map). It is now called Indian Ranch at the foot of the Hall Canyon alluvial fan. Young may have been unaware the horses were used for food, not for riding.

28. This was Townes's mess, with whom the Briers were recently traveling.

29. The peaks were Telescope (11,043 feet), Sentinel (9,634 feet), and Porter (9,101 feet). Panamint City was in Surprise Canyon, at the northwest base of Sentinel Peak. The lode was discovered in 1872, and by 1874 the town had two thousand residents.

30. Harry Vance is not listed in the various Jayhawker lists. Therefore, I list him under "Others" in Appendix B.

31. "Right East of your camp" may refer to the small fire the two Germans built to cook their supper, or a mess of Jayhawkers may also have stopped at the puddle water before continuing through the sand to camp at Borax Flat, still six miles west of south from Isham Canyon.

32. Goller (Galler) settled in Los Angeles. He later went back to prospect at least as far north as the El Paso Range.

Notes for Chapter 8

1. The mileage numeral in Young's diary is difficult to read. Colton's typist read it as "6," but we think it is more likely a "5" after looking at electronic scans of the original diary.

2. Great Falls Basin is also a contender for the honor of being called Providence Spring, but between the two we prefer Indian Joe Spring.

3. Brier mentions Vance before arriving in Searles Valley. Either there is faulty memory, or some of the men from the southern contingent caught up to the Briers in Panamint Valley.

4. Juliet's newspaper interview in 1898 implies St. John and Patrick were with the Briers at Post Office Spring. However, her 1905 letter is supported by Sheldon Young's diary (January 14) and James Brier's 1898 letter. It is possible the men arrived earlier but did not ask for help until Providence Spring.

5. Asa Haynes's group also included the Alonzo Clay mess and the four teamsters who abandoned the Bennett and Arcan families in Death Valley. "After leving Dotys camp, we camped with you the next night" (Manly 1894, 160).

6. On his "long hike" LeRoy tasted the water and declared, "Paxton Lake is salt" (January 20, 1973, 25). He also found suspected sheets of ice as Manly described.

7. Stephens comes close to mixing up the packers who left Mount Misery in Utah to follow the Smith and Flake packers and the Bug Smashers who parted from the Jayhawkers in Panamint Valley. At the Sheep Range in Nevada, eleven packers dried the flesh of their remaining horses and headed directly west for Walker Pass. Only Pinney and Savage are known to have arrived in California, and they told stories of impending cannibalism. This created rumor and supposition as to what happened to the remaining nine men. Thanks to William Lorton we have the names of these men (November 18 [19], 1849; January 20, 1850).

8. Manly uses different phrasing in *Death Valley in '49* (320–67), but he imparts the same information. We prefer the 1889 version unfiltered through the young editor of his book.

9. Haynes in various versions of his diary: "Campped at the foot of the serinavady mountains; thence South 15 miles—thence E of S 9 miles thence S 8 miles." This was in the vicinity of Bird Spring Canyon–Little Dixie Wash.

10. This was on their 250-mile hike from Travertine Springs in Death Valley to the site of the casa on the San Francisco Rancho near today's Castaic Junction.

11. There may have been water on (generally dry) Koehn Lake with grass nearby, or Young found the small playa where the Bennett and Arcan party camped a month later two miles southwest of the mouth of Last Chance Canyon. Manly says the playa was "about 4 miles" from Desert Spring, his "willow corral" (L. and J. Johnson 1987, 130).

12. These were the teamsters who left the Bennett and Arcan families in Death Valley and caught up to the Jayhawkers at Providence Spring. The assumption of many of the men was that the Panamint Range was really the Sierra Nevada, and once it was crossed, food was not far away. Thus, the teamsters were totally unprepared for the additional excruciating days in the desert.

13. Manly skews the directions on the map he sent to T. S. Palmer so he could squeeze it onto one page. The north-south ranges from the Funeral Mountains to the Sierra Nevada are turned west-east; the Mojave Desert is correctly south, but Manly draws the western route from Soledad Pass to the rancho as southward. He sometimes skews his cardinal directions in print as well. See more about the map in Appendix F.

14. *Snuff* was changed to *flour* by Manly's young female editor in *Death Valley in '49* (165), just as she morphed Manly's coyotes into wolves.

15. Tom Shannon and Ed Doty carried rifles out of the desert. Tom gave his to the California Historical Society, and Ed's is in the Curatorial Collection, Death Valley National Park (T. Shannon 1900). The Historical Society in San Francisco suffered severe damage in the 1906 earthquake and fire and has no record or relic of Shannon's rifle. Wolf Tauber either had a weapon or borrowed one to shoot a bullock at the Rancho San Francisco.

16. He says it was the day Robinson died (January 28) at or near Barrel Springs, but his portrayals of distances and the muddy pool are descriptive of Willow Springs. This was at least a day before Robinson died.

17. Young says the direction was "W of S," but Barrel Springs is east of south from Willow Spring. This mixing of directions is not uncommon; Manly does it, and we have too. At some point between the two springs, the Jayhawkers came across Manly and Rogers's tracks where they came into the Indian horse-thief trail.

18. Food scraps were put into a swill barrel to feed the pigs.

19. The editor of the *Santa Clara Valley* prefaced the January 1889 issue by saying Manly met with Doty, Shannon, and Stephens "within the last month" and "talked over the story of their travels."

20. The little brook is still flowing in the Acton area of Soledad Canyon.

Notes for Chapter 9

1. Chapter title quote is from Charles Mecum 1901.

2. This six miles brought them across Soledad Pass. Young does not mention the snow. A few junipers are often mixed in piñon forests, and these may be Young's "cedar."

3. Young's squiggles after "oak" appear to be "&c." The oak is coast live oak, *Quercus agrifolia*.

4. "Bolly" possibly refers to the white and tan color of bolls of cotton. It's written as "Colly Ox" in one copy. "Santa Clara Canyon" is today's Soledad Canyon, in which the Santa Clara River flows.

5. This would be Ellenbecker's man named "Coverly" (1993, 53) or T. S. Palmer's Jerome Bonaparte Aldrich (Wheat 1939a, 105).

6. This thumb of land at Saugus separates the main Santa Clara River from its South Fork, which waters today's city of Santa Clarita. The "city" incorporated Saugus, Valencia, Newhall, and Canyon Country into one large megopolis.

7. In the 1850s, the rancho was essentially in "probate" as Ignacio del Valle, one of the heirs, fought for the inheritors' rights through the entanglements of the U.S. legal system. Legally, they were

not yet "owners," but as inheritors they were responsible for management and profit sharing from the rancho. Ignacio chose the western part of the rancho known as Camulos for his share of the inheritance. His Rancho Camulos is now a National Historic Landmark near Piru.

8. Francisco's wife was Antonia, and the "daughters" were Indian women who worked in the casa.

9. Juliet Brier to "Fellow-Travelers," unlabeled newspaper article, possibly 1880, n.d., JCHL, Scrapbook 1.

10. LeRoy rescued adobe tiles from the debris bulldozed over the hill and deposited them in the Death Valley National Park Curatorial Center. This was after vandals shattered the floor tiles and razed the remaining walls looking for mission treasures. Perkins excavated the site between 1935 and 1937 before vandals ruined it.

11. Manly assumed the child was one of Antonia's children, but according to the 1850 Census, there were no children living at the casa. It was probably a child of one of the several Indians working at the casa.

12. Feliz was Jacoba's maiden surname. Her mother's maiden name was López. Two of Jacoba's López uncles worked on the rancho. Jacoba's second husband was José Salazar, whom she married in 1844.

13. Before Manly used the incorrect "San Francisquito" as the rancho name, none of the Death Valley '49ers used it. After his book was published, a few '49ers followed Manly's lead and started calling the rancho "San Francisquito." For an explanation of Manly's mistake, see "Myths," Appendix D.

14. These were members of the San Joaquin Company from whom the Briers had separated at Mount Misery in Utah when many of the wagons turned back to follow Hunt down the Old Spanish Trail.

15. Colton was quoted in a Galesburg newspaper [1897?] from the *Pioneer*, JCHL, Scrapbook 2. His 600 miles would be to the northern mines, but he went to the Merced and Sonora area, closer to 450 miles.

Notes for Chapter 10

1. The first state legislature convened on December 17, 1849, and adjourned on April 22, 1850. It functioned as a state governing body months before California was admitted to the Union on September 9, 1850 (H. Jones, 1950, 5–14).

2. Although Colton preferred the title of colonel, his rank in the Civil War was captain. In 1849–50 he was, however, a cocky teenager. The stature of his early exploits grew as Colton aged.

3. A Tom or Long Tom is a wooden rocker, similar to a cradle, with cleats in the bottom covered with a perforated metal sheet. Gold-bearing earth goes through the perforations, while larger material stays above the sheet and is thrown away. Gold settles between the riffle cleats. To operate a Long Tom requires several men and fast-running water to separate the materials.

4. See Ancestry.com—North America, Family Histories, 1500–2000—descendants of Quartermaster George Colton.

5. Colton sent a photograph of the daguerreotype to Stephens to use in his book. The surface of a daguerreotype is delicate and exceptionally prone to scratching, hence their elaborate cases. The process to make copies of a daguerreotype for resale in 1850 would have been very difficult. The copy in the Huntington Library is a photograph of the original, whereas the other early 1850s picture of Colton is an original daguerreotype still in its case. See the portraits.

6. By the time Manly and Rogers returned to Death Valley, the Wade family had left and the Bennett and Arcan families were preparing to leave.

7. This information about John Leveritt West comes from a clipping from an unknown newspaper article by W. L. Manly titled "Still Living" (referring to Lew West, not Manly). The date "May '96" is added in Colton's hand. Scrapbook 2:145, JCHL.

8. I have not determined if Mary Ann Robinson Doty was related to William Robinson, who died on the desert trek. U.S. Census, 1870, 1880.

9. Las Varas Canyon is eight miles west of the Santa Barbara Airport in Goleta and below Santa Ynez Peak.

10. Manly was forty-two years old when he married Mary Ann Woods of Woodbridge, California, on July 10, 1862 (marriage license, July 9, 1862. County Recorder, San Joaquin County, CA).

11. According to Blair Davenport, cultural resources manager, Death Valley National Park, in a personal communication, October 21, 2014. Doty's rifle was donated to the National Park Service about 1978. A quote from an interview with Doty's son Henry says, "It is a .44 calibre percussion cap rifle. On top of the octagonal barrel and back of the sight is stamped 'J. H. Holbrook'; on the silver plate at the side is inscribed 'George Coulcher'" (1938, 5–6).

12. It is possible Shannon's rifle still exists in a historical society other than the one in San Francisco, where everything was destroyed in the 1906 earthquake and fire—possibly San Jose or Santa Clara County.

13. There are two letters designated JA866—1897 and 1898. I differentiate them as JA866(1) and JA866(2).

14. Ed's older brother, Noyes Elsworth, married Cornelia Kellogg, Cordelia's twin sister. William Kellogg, the twins' brother, was part of Lorton's mess from Illinois to Salt Lake City.

15. See Haynes, Diary, JA1051; and *Historical Encyclopedia of Illinois and Knox County*, 1902, 910.

16. James Brier does not refer to the note for $115 he took from Ed Doty for "my 7 splendid oxen" that had been given to the Briers by the Bug Smashers. (Brier's "splendid" oxen were mostly skin and bones after their desert ordeal.) Brier sold the note to Dr. Earl (in Los Angeles?), thus acquiring additional capital (James Brier 1898).

17. Juliet Brier had to travel by boat from Santa Cruz to San Francisco in 1851 with three boys and a six-week-old baby when James was called to serve in Marysville (*Daily Alta Californian*, September 25, 1851, in Leadingham 1964, 122).

18. See Leadingham 1964, 122, 124; and U.S. Censuses 1850, 1860, 1870 for Santa Cruz, Santa Clara, and Solano Counties, CA. There were several French Camps in the gold country (Gudde 1975, 123).

19. See Lodi Memorial Cemetery Records; San Joaquin Co., Death Certificates Index; and John Brier 1884.

20. Either John Wells was assigned to preach temporarily in Santa Cruz, or Juliet was staying for a time with one of her daughters (*San Francisco Call*, February 9, 1913).

21. Manly wrote an article about Shannon for the *California Pioneer* and sent it to the 1894 Jayhawker reunion. See Manly in Woodward 1949, 45–48.

WORKS CITED

All letters referenced with a JA number are from the Jayhawkers of '49 Collection, the Huntington Library, San Marino, California, abbreviated as JCHL. Pages in Scrapbook 1 are not numbered. Some materials in Scrapbook 2 are loose between pages and have no page numbers. Scrapbook references also use the JCHL abbreviation. The location from which a letter was sent follows the date. Spelling of an author's or recipient's name has been standardized when different spellings were given. The James Brier letters to the Jayhawker reunions for 1876, '79, '93, '96, '98 are reproduced in LeRoy and Jean Johnson, eds., *Escape from Death Valley*, 176–87.

Aberdeen-Angus Journal. January 10, 1920. "Col. Colton, Death Valley '49er, Gone."

Allen, R. H. [George Allen's brother]. January 19, 1877. Chico, CA. Letter to John B. Colton. JA4.

Annals of Knox County. 1980. Galesburg, IL: Knox County Genealogical Society. Reproduction of the 1921 edition.

Annual Report of the Surveyor General of California. For 1855–56, 34th Cong., 3rd sess., Senate Ex. Doc. 5, 384–85. For 1856–57, 35th Cong. 1st Sess., House Ex. Doc. 2, 221, 224. *Map of Public Surveys in California to Accompany Report of Surveyor Genl.*, 1857.

Anthony, C. V. 1901. *Fifty Years of Methodism: A History of the Methodist Episcopal Church Within the Bounds of the California Annual Conference from 1847 to 1897*. San Francisco: Methodist Book Concern.

Arms, Cephas. 1849–50. "Cephas Arms Journal, 1849 May 20–1850 Jan. 23." *Knoxville (IL) Journal*, October 17–June 12. Dale Morgan typescript from the University of California, Berkeley, Bancroft Library, "Guide to the Dale Lowell Morgan Papers, 1877–1971." BANC MSS 71/161 c. Also in *The Long Road to California: The Journal of Cephas Arms Supplemented with Letters by Traveling Companions on the Overland Trail in 1849*. Edited by John Cumming. Mount Pleasant, MI: privately printed, 1985.

Ashby, Mrs. C. B. [William B. Rude's sister]. March 6, 1895. Letter to John Colton. JA9.

Bagley, Will. 2002. *Blood of the Prophets: Brigham Young and the Massacre at Mountain Meadow*. Norman: University of Oklahoma Press.

Bancroft, Hubert Howe. 1889. *The Works of Hubert Howe Bancroft*. Vol. 26, *History of Utah, 1540–1888*. San Francisco: History Publishing.

Bartholomew, A. R. [Edward's son]. January 10, 1889. Pueblo, CO. Letter to C. B. Mecum. JA19.

Bartholomew, E[dward] F[ranklin]. January 21, 1872. Pleasanton, KS. Letter to Colton. JA25.

———. January 18, 1875. Kansas. Letter to Dear Friend [Luther A. Richards]. JA33.

———. January 28, 1877. Pleasanton, KS. Letter to Colton. JA26.

———. January 6, 1878. Mound City, KS. Letter to Mr. Alonzo C. Clay. JA21.

———. January 11, 1879. Mound City, KS. Letter to Mr. Luther A. Richards. JA34.

———. January 23, 1880. Buena Vista, CO. Letter to Chs B. Mecomb [Mecum]. JA32.

———. January 26, 1881. Buena Vista, CO. Letter to Capt Asa Haynes. JA30.

———. January 16, 1884. Pueblo, CO. Letter to Mr. Luther A Richards. JA35.

———. January 22, 1885. Pueblo, CO. Letter to Capt Asa Haynes. JA31.

———. January 17, 1886. Pueblo, CO. Letter to Mr. Luther A Richards. JA36.

———. January 10, 1888. Pueblo, CO. Letter to Hon Alonzo C. Clay. JA23.

———. August 25, 1889. Pueblo, CO. Letter to John B. Colton. JA27.

———. January 22, 1890. Pueblo, CO. Letter to Mr. John B. Colton. JA28.

——. February 1, 1891. Pueblo, CO. Letter to Mr. C M [A. C.] Clay. JA24.

——. February 14, 1891. "E. F. Bartholomew Dead: He Died Suddenly at His Home of Heart Disease." *Pueblo (CO) Press*.

Bartholomew, George Wells, Jr. 1885. *Record of the Bartholomew Family: Historical, Genealogical and Biographical*. Austin, TX: privately printed.

Bartholomew, Lura H. 1905. "Capturing a Grizzly." *Western Field: The Sportsman's Magazine of the West* (San Francisco) 8 (4–5 May–June): 313–18.

Beaver Valley (NE) Tribune. February 4, 1894. "Jayhawker Reunion: A Gathering Reviving Thrilling Memories of Forty-nine. Awful Experience in Death Valley." JCHL, Scrapbook 1. (May have been *Beaver City [NE] Tribune*.)

——. February 11, 1898. "Survivors of the Perils of Death Valley: Annual Reunion of the Jayhawkers of '49 Held at the Home of Luther A. Richards Near Beaver City." JCHL, Scrapbook 2: 238.

Beck, John W., to Carl I. Wheat. May 21, 1939. Carl I. Wheat Collection, Bancroft Library, Berkeley, CA.

Belden, L. Burr. 1954. *Death Valley Heroine: And Source Accounts of the 1849 Travelers*. San Bernardino, CA: Inland Printing and Engraving.

Benson, Charles [son of Anderson Benson]. August 13, 1971. Letter to Death Valley Museum. "Discovery of Lost Gunsight Mine." Death Valley National Park Archives.

Berrett, LaMar C., and A. Gary Anderson. 2007. *Sacred Places: Wyoming and Utah, a Comprehensive Guide to Early LDS Historical Sites*. Vol. 6. Salt Lake City: Deseret Book.

Bidlack, Russell F. 1960. *Letters Home: The Story of Ann Arbor's Forty-Niners*. Ann Arbor, MI: Ann Arbor Publishers.

Bieber, Ralph P., ed. 1937. *Southern Trails to California in 1849*. Vol. 5, *The Southwest Historical Series*. Glendale, CA: Arthur H. Clark.

Bigler, David L., and Will Bagley, eds. 2000. *Army of Israel: Mormon Battalion Narratives*. Vol. 4, *Kingdom in the West: The Mormons and the American Frontier*. Spokane, WA: Arthur H. Clark.

Brier, Charles Templeton. 1956. *The Brier Family: A Collection of Data Tracing This Family from Its Earliest Beginnings in Dumfriesshire, Scotland to the Present Day in the New World*. Sacramento: privately printed.

Brier, James Welch. September 23, 1838. Record of marriage, James Brier to Juliet Wells. Centreville, MI, Circuit Court Records.

——. August 24, 1853. Pacific railroad meeting in San Francisco. In *Central Route to the Pacific by Gwinn Harris Heap*, edited by LeRoy R. Hafen and Ann W. Hafen, 277–81. Glendale, CA: Arthur H. Clark, 1957.

——. January 17, 1876. Grass Valley, CA. Letter to Charles B Mecum. JA70.

——. January 23, 1879. Glen Dale Farm, Nevada County, CA. Letter to Luther A. Richards. JA71.

——. December 21, 1886. "Death Valley: Its Ghastly Story as Told by an Aged Survivor: Rev. J. W. Brier, Who Preached the First Protestant Sermon in Los Angeles, Tells of That Awful Journey." *Los Angeles Times*, 4.

——. February 5, 1890. "The Jayhawkers: A Reunion of Old Forty-Niners." *Kansas City Times*. JCHL, Scrapbook 1.

——. January 10, 1893. Letter to John B. Colton. JA69.

——. January 16, 1896. Antiock, CA. Letter to Dear Old Comrades. JA76.

——. January 20, 1898. Lodi, CA. Letter to Mr. Deacon Richards. JA73.

Brier, John Wells. January 21, 1884. Salinas, CA. Letter to Luther A. Richards. JA100.

——. 1903. "The Death Valley Party of 1849." *Out West Magazine* 18 (April–May): 326–35, 456–65.

——. September 4, 1908. Lodi, CA. Letter to Colton. JA88.

——. 1911a. "The Argonauts of Death Valley." *Grizzly Bear* 9 (June): 1–4, 7.

——. 1911b. "A Lonely Trail." *Grizzly Bear* 9 (October): 1–2.

Brier, Juliet. February 25, 1849. Letter to My dear Brother. In Leadingham February 1964.

————. May 29, 1849. Letter to Parents and Brother. In Leadingham May 1964.

————. January 21, 1880. Letter to Our Fellow-Travelers. JCHL, Scrapbook 1.

————. December 25, 1898. "Our Christmas amid the Terrors of Death Valley: For the First Time Since 1850 Mrs. Julia Brier, the Only Woman of the Party, Relates the Awful Experiences of the Trip." *San Francisco Call*. Also in Belden 1954, 21–28.

————. January 13, 1902. Lodi, CA. Letter to My much respected friends, the Jay Hawks. JA104.

————. January 23, 1904. Letter to My fellow sufferers, survivors of the desert who entered refuges Feb 4. 1850. JA105.

————. January n.d., 1905. Lodi, CA. Letter to My respected Friends and Companions on our Journey to this Golden Shore. JA106.

————. January 23, 1906. Letter to the Friends and Companions who meet at the place appointed on the fourth of Feb 1906. JA107.

————. May 27, 1913. Obituary. *Sacramento Bee*.

————. June 8, 1913. "Death of Aged Woman Recalls First Trip Through Death Valley: Famous 'Jayhawker' Party the First to Struggle thru Sands of Deadly Desert." *Carson City News*. Original in *San Francisco Examiner*, February 24, 1901. Also in Belden 1954, 31–35.

Brockelbank, A. M. February 23, 1849. Oquawka, IL. Letter to John Colton's father [Chauncey Sill Colton]. JA108.

Brooks, Juanita. 1962. *The Mountain Meadows Massacre*. Norman: University of Oklahoma Press.

Brown, Randy. 2004. *Historic Inscriptions on Western Emigrant Trails*. Independence, MO: Oregon-California Trails Association.

Bruff, J. Goldsborough. 1944. *Gold Rush: The Journals, Drawings, and Other Papers of J. Goldsborough Bruff, Captain, Washington City and California Mining Association, April 2, 1849–July 20, 1851*. Edited by Georgia Willis Read and Ruth Gaines. 2 vols. New York: Columbia University Press.

Bullock, Thomas. 1997. *The Pioneer Camp of the Saints: The 1846 and 1847 Mormon Trail Journals of Thomas Bullock*. Edited by Will Bagley. Spokane, WA: Arthur H. Clark.

Burial Index-Sacramento Historic City Cemetery. oldcitycemetery.com.

Caesar, Ed. September 2014. "What Lies Beneath." *Smithsonian* 45 (5): 30–41.

California State Census, 1852. Online sources.

Cannon, George Q. 1869. "Twenty Years Ago: A Trip to California." *Juvenile Instructor* 4 (10): 78–79.

————. 1999. *The Journals of George Q. Cannon*. Vol. 1, *To California in '49*. Edited by Michael N. Landon. Salt Lake City: Deseret Book Co.

Carson, Kit. 1966. *Kit Carson's Autobiography*. Edited by Milo Milton Quaife. Lincoln: University of Nebraska Press. Reprinted from Lakeside Classics, Lakeside Press, no. 33. Chicago: R. R. Donnelley & Sons, 1935.

Caughey, John Walton, ed. See Stover.

Clay, Alonzo C. September 22, 1849. Great Salt Lake City. Letter to John T. Barnett, Esq. JA120.

————. January 7, 1882. Galesburg, IL. Invitation to Leander Woolsey. JA131.

————. January 1, 1888. Galesburg, IL. Invitation to the Reverend J. W. Brier and wife. JA121.

————. January 4, 1891. Galesburg, IL. Invitation to John L. West. JA129.

————. Obituary. [December 14, 1897]. "Death of Hon. A. C. Clay." JCHL, Scrapbook 2: 236. This extensive obituary probably came from one of the Galesburg papers.

Clayton, W. 1921. *William Clayton's Journal: A Daily Record of the Journey of the Original Company of "Mormon" Pioneers from Nauvoo, Illinois, to the Valley of the Great Salt Lake*. Edited by Lawrence Clayton. Salt Lake City: Deseret News.

Cole, Ruth. See Haynes n.d.

Colton, John B. April 15, 1849. Near Washington, IA. Letter to Dear Father [Chauncey Sill Colton]. JA176.

————. May 22, 1849a. Council Bluffs, IA. Letter to friends via James H. Notewan. JA250.

————. May 22, 1849b. At ferry above Kanesville, IA. Letter to friends via Emily Colton. JA179.

————. July 17, 1849. Letter to Dear Friends. "Camp on the Banks of the Sweet Water." *Oquawka (IL)*

Spectator, October 3, 1849. In Guide to the Dale Lowell Morgan Papers, typescript, microfilm pages 362–67, Bancroft Library, Berkeley, CA. First published in *Galesburg Gazetteer* about September 27, 1849. Also reprinted as "From on Route to California." *Monmouth Atlas*, October 5, 1849.

——. July 28, 1851. Sonora, CA. Letter to Dear Father. JA177.

——. September 11, 1851. San Francisco. Letter to Dear Father. JA178.

——. January 9, 1877a. Galesburg, IL. Invitation to Urban B. Davidson [Davison]. JA182.

——. January 9, 1877b. Galesburg, IL. Invitation to L. Dow Stephens. JA290.

——. January 8, 1883. Galesburg, IL. Invitation to John W. Plummer. JA263.

——. February ?, 1888. "Afoot in the Desert: Perils of a Party Who Crossed the Continent Back in '49." *Kansas City Times*. JCHL, Scrapbook 1.

——. January 9, 1890. Kansas City, MO. Invitation to John L. West. JA299.

——. January 10, 1890. Invitation to Brier & wife. JA160.

——. January 3, 1893. Kansas City, MO. Invitation to Rev. J. W. Brier & wife. JA162.

——. March 21, 1894. Kansas City, MO. Letter to T. S. Palmer. JA256.

——. April 12, 1894. Kansas City, MO. Letter to T. S. Palmer. JA257.

——. February 5, 1895. In "The Jayhawkers: A Terrible March." *Kansas City Journal*. JCHL, Scrapbook 2:174 continued from 170.

——. November 20, 1895. Letter to the editor. "The Death Valley Mines." *California Pioneer*, December 15, 1895. JCHL, Scrapbook 2.

——. [1897?]. Letter to the editor of the *California Pioneer*. Reprinted in a Galesburg newspaper as "A Jayhawkers' Reminiscences: John B. Colton Gives Interesting Facts About Early California." January 30, 1897. JCHL, Scrapbook 2: 151.

——. December 16, 1897. "John B. Colton's Estimate: Talks of His Friend and Comrade, the Late A. C. Clay, a Jayhawker." JCHL, Scrapbook 2: 237.

——. January 1, 1901. Letter to Dear Comrade. JA321.

——. 1904. Excerpt from undated letter in a footnote (p. 13) in Fox, S. M. 1904. "The Story of the Seventh Kansas." *Transactions of the Kansas State Historical Society, 1903–1904*, 8:13–49. Topeka, KS: Geo. A. Clark, State Printer.

——. May 1908. "The Jayhawkers of '49: Written by Col. John B. Colton." *Death Valley Magazine* 1 (7): 79–86. Subsequent issues (8–10) are attributed to editor Paul De Laney.

——. March 8–9, 1915. See Lockley 1915.

Colton, John Milton. 1912. *A Genealogical Record of the Descendants of Quartermaster George Colton*. Philadelphia: printed for private circulation. Collected and arranged for public use by George Woolworth, same title. Lancaster, PA: Wichersham, 1912.

Coolidge, Dane. 1937. *Death Valley Prospectors*. New York: E. P. Dutton.

Crampton, C. Gregory, and Steven K. Madsen. 1994. *In Search of the Spanish Trail: Santa Fe to Los Angeles, 1829–1848*. Salt Lake City: Gibbs-Smith.

Crespí, Juan. 2001. *A Description of Distant Roads: Original Journal of the First Expedition into California, 1769–1770*. Edited and translated by Alan K. Brown. San Diego: San Diego State University Press.

Cronkhite, Daniel. 1981. *Death Valley's Victims: A Descriptive Chronology, 1849–1980*. Morongo Valley, CA: Sagebrush Press.

Cumming, John. 1985. *The Long Road to California: The Journal of Cephas Arms Supplemented with Letters by Traveling Companions on the Overland Trail in 1849*. Edited by John Cumming. Mount Pleasant, MI: privately printed. See also Arms 1849–50.

Daily Illinois State Register (Springfield). October 27, 1919. "Col. Colton, One of 'Jayhawkers' Dead at Galesburg."

Daily Iowa Capital. [February] 1896. "Report of Jayhawk Meeting '96." JCHL, Scrapbook 2:136.

Daily Republican-Register (Galesburg, IL). February 4, 1895. "The Jayhawkers Meet: Talk over Terrible Experiences in Death's Valley." JCHL, Scrapbook 2:142.

Dame, William H. 1858. "Journal of the Southern Exploring Company for the Desert." Church Archives, Church of Jesus Christ of Latter-day Saints, Salt Lake City.

Davison, Urban P. January 25, 1878. Thermopolis, WY. Letter to Clay, JA337. Signed U.P.D.

———. December 28, 1883. Labarge, Sweetwater County, Wyoming Territory. Letter to Clay. JA360.

———. January 27, 1891. Letter to Elonzo Clay, Esq. JA338.

———. January 28, 1898. Thermopolis, WY. Letter to Luther A. Richards. JA358.

———. January 26, 1899. Thermopolis, WY. Letter to Colton. JA341.

De Decker, Mary. 1970. *The Eichbaum Toll Road.* Keepsake no. 10. Published for the dedication of the Eichbaum Toll Road Monument, Death Valley '49ers Encampment, November 13, 1970. San Bernardino, CA: Inland Printing and Engraving.

De Laney, Paul. 1908. "The Jayhawkers of '49: Written by Paul De Laney from Notes of Col. John B. Colton." Pts. 1–3. *Death Valley Magazine* 1 (8): 99–102; (9): 115–17; (10): 155–57. Rhyolite, NV. The first issue (1 [7]) was attributed to Colton.

Delta (Visalia, CA). April 13, 1861. "The Boundary Line—Face of the Country—Death Valley—Mineral Wealth," 2:2.

Dibblee, T. W., Jr. 1963. *Geology of the Willow Springs and Rosamond Quadrangles California.* USDI, Geological Survey Bulletin 1089-C. Washington, DC: GPO.

Dictionary of California Land Names. 1946. Compiled by Phil Townsend Hanna. Los Angeles: Automobile Club of Southern California.

Dobbs, David. 2013. "Restless Genes." *National Geographic* 223 (1): 44–57.

Doctor, Joseph E. 1988. *Dr. Samuel Gregg George: Death Valley Explorer of 1860–61.* Keepsake no. 28. Death Valley, CA: Death Valley '49ers.

Doty, Edward. April 2, 1849. "Last Will and Testament of Edward Doty, Box 877." Knox County, IL: Office of Records, SL/LDS, film #1403740.

———. January 21, 1872. San Jose, CA. Letter to Friend J. B. Colton. JA378.

———. January 18, 1873. San Jose, CA. Letter to A. C. Clay. JA381.

———. January 18, 1875. San Jose, CA. Letter to Friend Richards. JA391.

———. Postmarked January 10, 1876. Letter to Friend A. C. Clay. JA382.

———. January 25, 1882. Santa Barbara. Letter to Friend Clay. JA383.

———. January 24, 1883. Ever Green ranch. Letter to John B. Colton. JA 379.

———. January 25, 1884. 20 miles from Santa Barbara. Letter to Luther A. Richards. JA393.

———. January 23, 1886. Letter to Luther A. Richards. JA394.

———. January 12, 1887. Goleta, CA. Letter to Friend C. B. Mecum Esq. JA390.

———. January 13, 1888. Ever Green Ranch, Santa Barbara. Letter to Friend Clay. JA384.

———. January 28, 1890. Santa Barbara. Letter to Friend Colton. JA380.

Doty, Frank. January 23, 1892. Naples [Goleta], CA. Letter to Charles B. Mecum. JA396.

Doty, Henry. 1938. "Captain Edward Doty of the Death Valley Jayhawkers." Interview of Henry Doty, Edward Doty's son, by Olaf T. Hagen, July 19, 1938, at Buellton, CA. Typed, July 23, 1938, 7 pp. Includes a brief Edward Doty genealogy. T. S. Palmer Collection, Huntington Library, San Marino, CA.

Dubuque (IA) Tribune. August 10, 1849. "California Company."

Duerr, David. 2011–16. Personal communications via e-mail. Author files.

Editor. January 1908. "The Death Valley Magazine" 27 (1): 134.

Edwards, E. I. 1940. *The Valley Whose Name Is Death.* Pasadena, CA: San Pasqual Press.

———. 1964. *Freeman's, a State Stop on the Mojave.* Glendale, CA: La Siesta Press.

Ellenbecker, John G. 1993. *The Jayhawkers of Death Valley.* 2nd ed. Marysville, KA: printed by the author. Photo offprint of 1938 first edition plus his 1945 *Supplement to the Jayhawkers of Death Valley.* The 1993 edition includes both the above.

Emigrant Rosters, 1849. MS 15494. Salt Lake City: LDS Church, Family and Church History Department. Handwritten, dated, lists of emigrant arrivals.

Evans, George W. B. 1945. *Mexican Gold Trail: The Journal of a Forty-Niner*. Edited by Glenn S. Dumke. San Marino, CA: Huntington Library Press.

Farley, Minard H. 1861. "Farley's Map of the Newly Discovered Tramontane Silver Mines in Southern California and Western New Mexico [now Nevada]." San Francisco: W. Holt. This is probably the first published map with "Death Valley" and "Town's Pass" designated.

Feynman, Richard P. 1999. *The Pleasure of Finding Things Out*. Edited by Jeffrey Robbins. Cambridge, MA: Perseus Books.

Findley, William R. August 16, 1849. "Gold Diggings, 40 Miles East of Johnson's." *Oquawka (IL) Spectator*, October 31, 1849.

Fish, Jeremiah 1850. "Letter from J. Fish, Mission of San Gabriel, Cala., March 1st, 1850." *Iowa Democratic Enquirer* (Muscatine), June 1, 1.

Fox, S[imeon]. M. 1904. "The Story of the Seventh Kansas." In *Transactions of the Kansas State Historical Society, 1903–1904*, 8:13–49. Topeka, KS: Geo. A. Clark, State Printer.

Francaviglia, Richard V. 2005. *Mapping and Imagination in the Great Basin: A Cartographic History*. Reno: University of Nevada Press.

Frans, Harrison B. July 21, 1849. "Letter [to Dear Parents]." In *Oquawka (IL) Spectator*, October 10, 1849. Reprinted in *Galesburg (IL) Gazetteer* [October 1849]. Typescript in "Guide to the Dale Lowell Morgan Papers." BANC MSS 71/161 c, Bancroft Library, Berkeley, CA.

———. January 28, 1889. Baker City, OR. Letter to the Jayhawkers. JA449.

———. January 26, 1891. Baker City, OR. Letter to D[ea]r friend A. C. Clay and Jayhawks. JA434.

———. January 26, 1892. Rye Valley, Baker Co., OR. Letter to Charles B. Mecum. JA445.

———. March 22, 1892. Rye Valley, OR. Letter to Friend John B. Colton. JA440.

———. January 28, 1895. Baker County, OR. Letter to Alonzo C. Clay. JA435.

———. January 25, 1896. Baker, OR. Letter to Charles B. Mecum. JA446.

———. January 20, 1900. Rye Valley, OR. Letter to Mr. J. B. Colton. JA442.

———. October 8, 1901. Rye Valley, OR. Letter to Friend John B. Colton. JA443.

Franzwa, Gregory M. 1982. *Maps of the California Trail*. Tucson, AZ: Patrice Press.

Frémont, John Charles. 1970. *The Expeditions of John Charles Frémont*. Vol. 1, *Travels from 1838 to 1844*. Edited by Donald Jackson and Mary Lee Spence. Urbana: University of Illinois Press.

Frontier Guardian (Kanesville, IA). May 30, 1849. "Constitution and By-Laws of the Knox County Company, Illinois." Includes roster of names as of May 22, 1849. See also Wheat 1940.

Gale, W. Selden, and Geo. Candee Gale, eds. 1899. "Asa Haynes." *Knox County*. In *Historical Encyclopedia of Illinois and Knox County*, edited by Newton Bateman and Paul Selby. Chicago: Munsell.

Gentry, Leland H. 1981. "The Mormon Way Stations: Garden Grove and Mt. Pisgah." *Brigham Young University Studies* 21 (4): 445–61.

George, S. G. March 1, 1875. Letter to "Editor News." In "Our Early History." *Panamint News* (Panamint City, CA). Reprinted in C. Lorin Ray, "Letter Tells of Early Inyo Events." *Inyo Register* (Bishop, CA), December 29, 1966.

Gibbes, C. D. See Map 1852.

Gordon, James W., ed. 1911. *History of Henderson County*. In *Historical Encyclopedia of Illinois*. Edited by Newton Bateman and Paul Selby. Vol. 2. Chicago: Munsell.

Granger, Lewis. April 8, 1850. Letter to "Father." In *Letters of Lewis Granger: Reports of the Journey from Salt Lake to Los Angeles in 1849, and of Conditions in Southern California in the Early Fifties*. Edited by LeRoy R. Hafen. Los Angeles: Glen Dawson, 1959.

Grayson, Donald K. 1993. *The Desert's Past: A Natural Prehistory of the Great Basin*. Washington, DC: Smithsonian Institution Press.

Gritzner, Frederick. 1850. Letter to wife. In "Dreadful Suffering on the Plains." *Deseret News*, August 17, 1849. The article preceding the letter spelled Frederick's name as "Kritzner."

Groscup, John. February 20, 1872. Long Valley, CA. Letter to Mr. Colton. JA472.

———. n.d., 1875. Long Valley, CA. Letter to Mr. Richard. JA502.

———. January 18, 1885. Long Valley, [Cahto] CA. Letter to Mr. Captain Hanes and Jay Hawkers of 1849. JA498.

———. December 14, 1903. Laytonville, CA. Letter to Dear Comrade [John Colton]. JA477.

———. April 23, 1904. Laytonville, Mendocino Co. Letter to Mr. John B. Colton. JA481.

Gudde, Erwin G. 1975. *California Gold Camps: A Geographical and Historical Dictionary*. Berkeley: University of California Press.

Hafen, LeRoy R., ed. 1969. *The Mountain Men and the Fur Trade of the Far West*. Vol. 7. Glendale, CA: Arthur H. Clark.

Hafen, LeRoy R., and Ann W. Hafen, eds. 1954. *Journals of Forty-Niners: With Diaries and Contemporary Records of Sheldon Young, James S. Brown, Jacob Y. Stover, Charles C. Rich, Henry W. Bigler, and Others*. Glendale, CA: Arthur H. Clark.

———. 1961. *The Far West and Rockies: General Analytical Index to the Fifteen Volume Series and Supplement to the Journals of Forty-Niners, Salt Lake to Los Angeles*. Glendale, CA: Arthur H. Clark.

Hafen, LeRoy R., and Francis Marion Young. 1984. *Fort Laramie and the Pageant of the West, 1854–1890*. Lincoln, NB: University of Nebraska Press.

Hale, J[ames] Ellery. January 8, 1850. Bodega, CA. Letter to My Dear Parents. Reprinted in *Knoxville (IL) Journal*, April 10, 1850.

Hanks, Henry G. 1883. *Report on the Borax Deposits of California and Nevada*. Part. 2, *Third Annual Report of the State Mineralogist*. California State Mining Bureau. Sacramento: Superintendent State Printing.

Hanna, Phil. See *Dictionary of California Land Names*.

Harline, Jo Lynne [professional genealogist]. June 27, 2008. Personal communication. Author files.

Hartley, William G., and A. Gary Anderson. 2006. *Sacred Places: Iowa and Nebraska, a Comprehensive Guide to Early LDS Historical Sites*. Vol. 5, *Sacred Places*. Salt Lake City: Deseret Book.

Haynes, Asa. N.d. "The Story of Capt. Haynes." "The following is the story of Capt Haynes as he Dictated same to Nancy J. Wiley, his Daughter & later Re-written by W. A. Wiley's daughter Ruth Cole [Nancy J. Wiley's granddaughter]." JCHL, FAC 705, box 9, folder 44, 19 pp. See Wiley, Nancy J., for another form of her notes from Asa Haynes's accounts.

———. 1849–50. [Diary] "Knoxville Knox C[ounty]/April the 5, 1849 [Diary of] Asa Haynes." JA1051.

———. January 1, 1881. Knoxville, IL. Letter to Urban P. Davidson [sic]. JA513.

———. January 24, 1886. Delong, IL. Letter to Richards. JA514.

Heap, Gwinn Harris. 1854. *Central Route to the Pacific, from the Valley of the Mississippi to California: Journal of the Expedition…from Missouri to California, in 1853*. Philadelphia: Lippincott, Crambo, and Co. In *Central Route to the Pacific by Gwinn Harris Heap*, edited by LeRoy R. Hafen and Ann W. Hafen, 75–281. Glendale, CA: Arthur H. Clark, 1957.

Heitman, Francis B. 1903. *Historical Register and Dictionary of the United States Army*. Vol. 1. Washington, DC: GPO.

Hermann, William H. 1959. "Three Gold Rush Letters of Adonijah Strong Welch." *Iowa Journal of History* 57 (1): 61–73.

Hileman, Levida. 2001. *In Tar and Paint and Stone: The Inscriptions at Independence Rock and Devil's Gate*. Glendo, WY: High Plains Press.

Historical Encyclopedia of Illinois and History of Peoria County. 1902. Edited by David McCulloch. Vol. 2. Chicago: Munsell.

History of Buchanan County, Missouri, Containing a History of the County, Its Cities, Towns, Etc. 1881. St. Joseph, MO: Union Historical.

History of Johnson County, Iowa, Containing a History of the County, and Its Townships, Cities and Villages from 1836 to 1882. 1883. Iowa City: Iowa City.

History of Peoria County Illinois. 1880. Chicago: Johnson.

History of Tulare County, California…[with] Biographical Sketches. 1883. San Francisco: Wallace W. Elliott and Company.

Hopkins, Timothy. 1903. *The Kelloggs in the Old World and the New*. Vol. 1. San Francisco: Sunset Press and Photo Engraving.

Horst, Bill. 1996. "The Original Walker Pass." In *Proceedings: Fourth Death Valley Conference on History and Prehistory February 2–5, 1995*, edited by Jean Johnson, 16–24. Death Valley, CA: Death Valley Natural History Association.

Hoshide, Robert Kenn. 1996. "The Mississippi Boys, 1849–50." In *Proceedings: Fourth Death Valley Conference on History and Prehistory, February 2–5, 1995*, edited by Jean Johnson, 208–20. Death Valley, CA: Death Valley Natural History Association.

Hunt, Alice. 1960. *Archeology of the Death Valley Salt Pan California*. Anthropological Papers, no. 47. Salt Lake City: University of Utah Press.

Hunt, Alice P., and Charles B. Hunt. 1964. "Archaeology of the Ash Meadows Quadrangle, California and Nevada." Manuscript on file at Death Valley National Monument.

Inyo Independent (Independence, CA). October 4, 1884. "The Gunsight Mine" 2:1–3.

Jensen, Marvin. 2007. *The Remarkable Life of 1849 Death Valley Pioneer John B. Colton*. Keepsake no. 47. Death Valley, CA: Death Valley '49ers.

Johnson, Craig. 2015. *Dry Bones*. New York: Penguin Random House.

Johnson, Jean, ed. and comp. 2008. *Tall Tales of Death Valley*. Bishop, CA: privately printed.

———. 2018. "Francisco López: 'Verily, He Was a Good Samaritan.'" In *Proceedings: Tenth Death Valley Conference on History and Prehistory, November 5–8, 2015*, 396–407. Death Valley, CA: Death Valley Natural History Association.

Johnson, Jean and LeRoy. 2013. "Rancho San Francisco's Mystery Lady." In *Proceedings: Ninth Death Valley Conference on History and Prehistory, November 4–6, 2011*, 25–33. Death Valley, CA: Death Valley Natural History Association.

———. 2017. "Adonijah Strong Welch, 1849er." *Overland Journal* 35 (1): 22–30.

Johnson, Jean and LeRoy, with Ted Faye and Robert Ryan. 1999. *John Rogers: Death Valley's Unsung Hero of 1849*. Keepsake no. 39. Death Valley, CA: Death Valley '49ers, Inc.

Johnson, LeRoy C. January 1973–November 1975. "Diaries, Vol. 2." 180 pp.

———. 1999. "The Trunk Is Bunk: The Latest, Notorious Death Valley Artifacts." In *Proceedings: Fifth Death Valley Conference on History and Prehistory, March 4–7, 1999*, edited by Jean Johnson, 252–77. Bishop, CA: Community.

———. 2005. "Bennett's Trail into the Great Basin." In *Nuggets & Snippets of Death Valley History, Proceedings: Seventh Death Valley Conference on History and Prehistory, February 3–6, 2005*, edited by Jean Johnson, 43–74. Bishop, CA: Community.

———. 2013. "The Enigma of William B. Rude." In *Proceedings: Ninth Death Valley Conference on History and Prehistory, Temptations, Practicalities and Life, November 4–6, 2011*, 284–95. Death Valley, CA: Death Valley Natural History Association.

Johnson, LeRoy and Jean, eds. 1987. *Escape from Death Valley, as Told by William Lewis Manly and Other '49ers*. Reno: University of Nevada Press.

———. 1988. "Discovery of the 1861 Boundary Reconnaissance Map and Field Notebooks." In *Exploring Death Valley History: Proceedings First & Second Death Valley Conference on History and Prehistory*, edited by Jean Johnson, 147–60. Bishop, CA: Community.

———. 1992. "Where Is Van Dorn's 'Hitchins' Spring'"? In *Proceedings: Third Death Valley Conference on History and Prehistory, January 30–February 2, 1992*, edited by James Pisarowicz, 45–56. Death Valley, CA: Death Valley Natural History Association.

———. 1995. "Where Is Providence Spring'?" In *Proceedings: Fourth Death Valley Conference on History and Prehistory, February 2–5, 1995*, edited by Jean Johnson, 25–43. Death Valley, CA: Death Valley Natural History Association.

———. 1996. *Dr. J. R. N. Owen Frontier Doctor and Leader of Death Valley's Camel Caravan*. Death Valley: Death Valley '49ers.

———. 2007. "Discovery of the 1861 Boundary Reconnaissance Map and Field Notebooks." In

Exploring Death Valley History: Proceedings: First & Second Death Valley Conference on History and Prehistory, February 8–11, 1987, & January 21–25, 1988, edited by Jean Johnson, 147–60. Bishop, CA: Community.

———. 2013. *Julia, Death Valley's Youngest Victim: The Heroic Rescue of the Stranded 1849ers*. 3rd ed. Bishop, CA: Community.

———. 2014. "Porter Rockwell's Route Home." *Spanish Traces* (Winter): 20–23, 27–31.

———. 2018. "Where Is Mount Misery?" In *Proceedings: Tenth Death Valley Conference on History and Prehistory, November 5–8, 2015*, 10–43. Death Valley, CA: Death Valley Natural History Association.

Jones, Gurley. April 6, 1903. Letter to Mr. J. B. Colton. JA554.

Jones, Herbert C. 1950. *The First Legislature of California*. Address before California Historical Society, San Jose, December 10, 1949. California Senate, Paper 237.

Kane, Michael David. 2008. "William Lewis Manly." PhD diss., University of Utah.

Kansas City Times. [February] 1888. [Colton, J. B.] ?"Afoot in the Desert: Perils of a Party Who Crossed the Continent Back in '49." JCHL, Scrapbook 1.

———. February 5, 1890. "The Jayhawkers: A Reunion of Old Forty-Niners: Four Survivors of an Historical Party Grasp Hands." JCHL, Scrapbook 1:25, 29.

Kelly, Charles. February 1939. "On Manly's Trail to Death Valley." *Desert Magazine*, 2 (9): 4, 6–8, 41, 43.

Kimball, Stanley B. 1988. *Historic Sites and Markers Along the Mormon and Other Great Western Trails*. Urbana: University of Illinois Press.

Knox (or *Knox County*) *Republican*. See *Republican Register*.

Koenig, George. 1984. *Beyond This Place There Be Dragons: The Routes of the Tragic Trek of the Death Valley 1849ers Through Nevada, Death Valley, and on to Southern California*. Glendale, CA: Arthur H. Clark.

Kritzner, Fred. See Gritzner, Frederick.

Leadingham, Grace. 1963–64. "Juliet Wells Brier: Heroine of Death Valley." Pts. 1–3. *Pacific Historian* 7 (November 1963): 170–78; 8 (February 1964): 13–20; 8 (May 1964): 61–74.

Leonard, Zenas. 1978. *Narrative of the Adventures of Zenas Leonard: Written by Himself*. Edited by Milo Milton Quaife. Facsimile of first edition (1934). Lincoln, NB: University of Nebraska Press.

Levy, Benjamin. April 15, 1969. *Death Valley National Monument Historical Background Study*. Base Map No. 1. USDA, Division of History, Office of Archeology and Historic Preservation.

Lewin, Jacqueline A., and Marilyn S. Taylor. 1992. *The St. Joe Road*. St. Joseph, MO: St. Joseph Museum.

Lingenfelter, Richard E. 1986. *Death Valley & the Amargosa: A Land of Illusion*. Berkeley: University of California Press.

Lockley, Fred. March 8–9, 1915. "The Oregon Country 'in Early Days.'" *Oregon Daily Journal*, March 8, 4, col. 8, and March 9, 6, col. 8.

Lodi Memorial Cemetery, book 1. San Joaquin County, CA, Cemetery Records. F. H. L. Microfilm #1036000.

Long, Margaret. 1941. *The Shadow of the Arrow*. Caldwell, ID: Caxton.

———. 1950. *The Shadow of the Arrow*. 2nd ed. Caldwell, ID: Caxton.

Lopez, Pedro. "Testimony Before the Title Board of Commissioners about the Boundaries of the Rancho San Francisco, 1854." Containing 48,813 58/100 Acres, Scale 80 Cha[ins] to 1 inch." Folsom, CA. Bureau of Land Management, Docket 316, Roll 23, pp. 0501–0510. Copy in Bancroft Library and author's files.

Lorton, William B. September 1848–January 1850. "California Journals." Bancroft Library, University of California, Berkeley, CA, BANC MSS C–F 190, Vols. 1–9: William B. Lorton diaries and papers, 1848 September–1850 January.

———. September 1848–January 1850. "California Journals." Typescript translation by Dale L. Morgan. The Bancroft Library, Univ. of California, Berkeley, CA, BANC MSS 71/161 c.

———. January 20, 1850. Letter to the editor. "Correspondence of the New York Sun: Overland Emigration—Terrible Sufferings of the Emigrants—Thrilling Details." Written from "De Los Angelos." *New York Sun*, May 8, 1850. Reprinted as "Wanderings and Sufferings of a Party in the Great Basin." *New-York Daily Tribune*, May 17, 1850. Bancroft Library, University of California, Berkeley, BANC MSS C–F190, Vols. 1–9, William B. Lorton diaries and papers, September 1848–January 1850.

———. [Forthcoming]. *Troubadour on the Road to Gold: William B. Lorton's Journals to California, 1849*. Edited by LeRoy and Jean Johnson. Cedar City: Southern Utah University Press.

Los Angeles Star, November 24, 1860. "Arrivals at the Bella Union Hotel, for the Week ending...", p. 2:3. [Concurrent arrival of Manly, Stockton, and Brooks.]

Lyman, Edward Leo. 2004. *The Overland Journey from Utah to California: Wagon Travel from the City of Saints to the City of Angels*. Reno: University of Nevada Press.

MacBrair-Koller, E. B. May 2, 1988. Letter to authors. Johnson files.

Madsen, Brigham D. 1982. *A Forty-Niner in Utah: Letters and Journal of John Hudson*. Salt Lake City: Univ. of Utah Library.

———. 1983. "The Colony Guard: To California in '49." *Utah Historical Quarterly* (Winter): 5–28.

Madsen, David B. 1989. *Exploring the Fremont*. University of Utah Occasional Paper, no. 8. Salt Lake City: Utah Museum of Natural History.

Manly, W[illiam]. L[ewis]. August 23, 1884. "A Survivor of the Death Valley Tragedy." Letter to the editor. *Inyo Independent*.

———. June 1887–July 1889. "From Vermont to California." *Santa Clara Valley: Pacific Tree and Vine* (later called *California Country Journal*) 4 (4) through 7 (5). These thirty-eight articles were bylined "Manley."

———. February 27, 1888. College Park, San Jose, CA. Letter to Friend Clay. JA613.

———. January 27, 1889. College Park, San Jose. Letter to My good friends. JA644.

———. ca. 1889. Letter to John Colton. Includes map of Jayhawkers' route south of Salt Lake to Rancho San Francisco. JA1050.

———. 1890. Map. "Showing the Trail the Emigrants Travailed from Salt Lake to San Bernardino [sic] in 1849." Sent to T. S. Palmer in August 1894. See Map 1890.

———. February 7, 1890. College Park, CA. Letter to John B. Colton. JA616. Manly sends a sketch map of the 1849 trails to Colton.

———. March 16, 1890. College Park, CA. Letter to J. B. Colton. JA617.

———. November 23, 1890. College Park, CA. Letter to A. C. Clay. JA614.

———. January 8, 1891. College Park, CA. Letter to Dear Jayhawkers (sent to Colton). JA645.

———. May 1892. San Jose, CA. Letter to Friend Colton. JA620.

———. January 10, 1893. San Jose, CA. Letter to Bro[ther] Jayhawkers. JA646[a]. (Manly wrote "Feb. 10, 1893" on the letter, but written above the date in what looks like Colton's hand is "Jany.")

———. October 15, 1893. "A Relic of '49." JCHL, Scrapbook 2, 42. Also in Woodward 1949, 45–48.

———. 1894. *Death Valley in '49*. San Jose, CA: Pacific Tree and Vine. Original editions of this book are rare, but it is available in several facsimile editions.

———. July 1894. College Park, CA. Letter to Friend Colton. JA625.

———. August 1894. College Park, CA. Letter to Friend Palmer. Palmer Collection, Huntington Library, San Marino, CA, HM59802. Included map Manly drew in 1890.

———. November 25, 1894. College Park, CA. Letter to A. C. Clay. JA615.

———. December 1894. College Park, CA. Letter to Friend Palmer. Palmer Collection. Huntington Library, San Marino, CA, HM 59803. (Manly stated the map he sent Palmer in 1894 was drawn 1890.)

———. May 1896. "Still Living." JCHL, Scrapbook 2:148.

———. December 1896. "Georgetown in Fifty." JCHL, Scrapbook 2 [near p. 183].

———. 1949. *The Jayhawkers' Oath, and Other Sketches*. See Woodward.

———. 1987. In *Escape from Death Valley as Told by William Lewis Manly and Other '49ers*. See L. and J. Johnson 1987.

Map. 1814. Allen, Paul, et al. "A Map of Lewis and Clark's Track, Across the Western Portion of North America From the Mississippi to the Pacific Ocean." Copied by Samuel Lewis from the Original Drawing of Wm Clark. Philadelphia: Bradford and Inskeep. In David Rumsey Historical Map Collection, www.davidrumsey.com.

Map. 1846. Mitchell, Samuel Augustus. "A New Map of Texas, Oregon, and California with the Regions Adjoining." Pocket map. Philadelphia: S. A. Mitchell. David Rumsey Historical Map Collection, www.davidrumsey.com.

Map. 1848. Frémont, John Charles. 1848. "Map of Oregon and Upper California: From the Surveys of John Charles Frémont And other Authorities. Drawn by Charles Preuss Under the Order of the Senate of the United States, Washington City, 1848." Map 5 in "Map Portfolio" of *The Expeditions of John Charles Frémont*, edited by Donald Jackson and Mary Lee Spence. Urbana: University of Illinois Press, 1970.

Map. 1849. Manly. Route from Salt Lake City to Rancho San Francisco, Santa Clarita Valley, CA. Drawn in 1890 by Manly for his friends. Sent to T. S. Palmer in 1894.

Map. 1849. Mitchell, Samuel Augustus. "Oregon, Upper California & New Mexico." S. A. Mitchell, Philadelphia. David Rumsey Historical Map Collection, www.davidrumsey.com.

Map. 1851. Henn, Williams & Co. "Township Map of the State of Iowa." Henn, Williams & Co., J. F. Abrahams, Burlington, Iowa. David Rumsey Historical Map Collection, www.davidrumsey.com.

Map. 1852. Gibbes, Charles Drayton. "A New Map of California." Stockton, CA: C. D. Gibbes and Sherman & Smith, New York. Online, Ramsey Collection. "Early state map, inaccurate yet elegant." Carl I Wheat: "This was the most elaborate map of California that had yet appeared. It is beautifully designed and drawn, . . . though it cannot be entirely commended for accuracy."

Map. 1858. Martineau, James H. "Chart Showing the Explorations of the Desert Mission." Church Archives, Church of Jesus Christ of Latter-day Saints, Salt Lake City.

Map. [1861]. [Van Dorn, Aaron]. "Pencil sketches [two maps] showing the Tulare & Death Valleys, CA. With adjacent country & line of a route beginning at camp 10 near Amargosa Butte & extending to Visalia past Owens Lake through Walker's Pass & via Keysville & Sextonville." National Archives. Record Group 77. Civil Works Map File. Map U.S. 324-121. Two sheets each 88.7 cm wide, 56.5 cm high.

Map. 1877. Wheeler, George M. Issued May 7. U.S. Geographical Surveys West of the 100th Meridian. Part of Eastern California, Atlas Sheet No. 65 (D.). [Washington, DC: Secretary of War.] See also Wheeler 1877.

Map. 1890. Manly, William Lewis. "Showing the Trail the Emigrants Travailed from Salt Lake to San Bernardino [sic] in 1849." Route from Salt Lake City to Rancho San Francisco, Santa Clarita Valley, CA. Manly sent map to T. S. Palmer in 1894. Palmer Collection, Huntington Library, San Marino, CA, HM 50895.

Map. 2009. National Geographic Maps. "TOPO! Outdoor Recreational Mapping Software." Utah, Nevada, and California, edition 4.5.4.

Marius, Richard. 1999. *A Short Guide to Writing About History*. 3rd ed. New York: Longman.

Martineau, James H. 1928. "A Tragedy of the Desert." *Improvement Era* 31 (July): 271–72. [Editor's note says: "This article was written in 1910. Since then its author has died, having passed away June 24, 1921."]

———. See Santiago, Martineau's pen name.

Marwitt, John. 1986. "Fremont Cultures." In *Great Basin*, vol. 2 of *Handbook of North American Indians*, edited by Warren L. l'Azevedo, 161–72. Washington, DC: Smithsonian Institution.

Mattes, Merrill J. 1969. *The Great Platte River Road*. Nebraska State Historical Society Publications, vol. 25. [Lincoln]: Nebraska State Historical Society.

McGilligan, Patrick. 1999. *Clint: The Life and Legend*. New York: St. Martin's Press.

McGrew, Miletus S. [Son of Thomas] January 18, 1872. Bloomington, IL. Letter to John B. Colton. JA606.

McGowan, Edward. Early 1872. Postcard to J. B. Colton. JA746.

Mecum, Charles B. March 20, 1851. Bidwell's Bar, Feather River, CA. Letter to Friend Barnett. JA665.

———, alias big paddy. January n.d., 1872. Letter to Honard Jay Hawkers and Shinpickers. JA715.

———. January 26, 1878. Rippey, IA. Letter to A. C. Clay, Dear Brother Jayhawkers. JA716.

———, Chief of the Piutes. February 13, 1880. Rippey, IA. Letter to Friend Colton. JA667.

———. January 6, 1892. Rippey, IA. Invitation to Rev. J. W. Brier & wife. JA666.

———. January 16, 1901. Independence, IA. Letter to L. D. Stevens. JA712.

Mecum, Edwin W. February 18, 1908. Camarillo, CA. Letter to John B. Colton. JA720.

———. January 17, 1910. Camarillo, CA. Letter to Mr. John E.[sic] Colton. JA722.

———. January 31, 1910. Camarillo, CA. Letter to John B. Colton. JA723.

———. April 12, 1931. "Report of the Jayhawker [descendant] Reunion of April 12, 1931." JCHL, Box 9, Folder 42.

———. October 5, 1931. "Biographies—Families: Charles Burt Mecum of the Jayhawkers." Knox County, Illinois Genealogy and History. http://knox.illinoisgenweb.org/biographies/bio_mecum_cb.htm.

Mecum, Frances E. March 24, 1889. Rippey, IA. Letter to Colton, "The 1st Book of Chronicles of the Jayhawkers." JA738.

———. February 23, 1900. Richland Eagle Valley? Letter to J. B. Colton. JA730.

———, F E M. February 21, 1905. Mt. Vernon, IA. Postcard to J. B. Colton. JA734.

Mecum, William F. February 9, 1917. Douglas, WY. Letter to Colton. JA744.

Meldahl, Keith Heyer. 2007. *Hard Road West: History & Geology Along the Gold Rush Trail*. Chicago: University of Chicago Press.

Mendenhall, Walter C. 1909. *Some Desert Watering Places in Southeastern California and Southwestern Nevada*. Water-Supply Paper 224. U.S. Geological Survey. Washington, DC: GPO.

Miller, George. 1919. "A Trip to Death Valley." *Historical Society of Southern California, Annual Publications* 11 (2): 56–64.

Mitchell. See Map, Mitchell.

Monmouth Atlas. October 5, 1849. "From on Route to California."

Montgomery, Lorenzo Dow. January 26, 1890. Denver. Letter to John B. Colton. JA1090.

Morgan, Dale L. 1959. "The Ferries of the Forty-Niners." *Annals of Wyoming* 31 (1): 4–31.

Munro-Fraser, J. P. 1881. *History of Santa Clara County, California*. San Francisco: Alley, Bowen.

Nelson, Genne. 1999. "The Personal Side of the Jayhawkers: Vignettes from the Jayhawker Collection, Huntington Library [San Marino, CA]." In *Proceedings: Fifth Death Valley Conference on History and Prehistory, March 4–7*, edited by Jean Johnson, 19–41. Bishop, CA: Community Printing and Publishing.

———. 2002. "Crossing the Plains in '49: The Trek of the Death Valley '49ers." In *Death Valley History Revealed: Proceedings Sixth Death Valley Conference on History and Prehistory, February 7–10*, edited by Jean Johnson, 246–82. Bishop, CA: Community Printing and Publishing.

Newmark, Maurice H. and Marco R. Newmark. 1929. *Census of the City and County of Los Angeles, California for the Year 1850*. Los Angeles: Times-Mirror Press.

New York Herald Tribune. November 28, 1848. "The Gold Mines in California," 2:2.

New York World. September 16, 1894. "The Horrors of Death Valley: Reptile-Inhabited, Sun-Baked, Waterless, Desolate Desert."

Northrop, Marie E. 1984. *Spanish-Mexican Families of Early California: 1769–1850*. Vol. 2. 2nd ed. Burbank, CA: Southern California Genealogical Society.

———. 1987. *Spanish-Mexican Families of Early California: 1769–1850*. Vol. 1. 2nd ed. Burbank: Southern California Genealogical Society.

Norway, W. H. 1877. "Part of R. San Francisco (Jacoba Felix et al.) Lot No. 37." T4N, R16W. Bureau of Land Management, Folsom, CA, ref. #52 276. Approved by the California Surveyor General,

H. G. Rollins, November 3, 1906. Johnson files. Norway surveyed the eastern portion of the Rancho San Francisco in September 1876. The map was approved on March 26, 1877.

Nusbaumer, Louis. 1849. Diary. In *Escape from Death Valley*, edited by LeRoy and Jean Johnson, 160–68. Reno: University of Nevada Press, 1987.

———. 1967. *Valley of Salt, Memories of Wine: A Journal of Death Valley, 1849.* Edited by George Koenig. Berkeley, CA: Friends of the Bancroft Library.

Nystrom, Alfred, City Clerk. September 30, 1936. Galesburg, IL. To T. S. Palmer. Palmer Collection, Huntington Library, San Marino, CA, HM 50808.

Oquawka (IL) Spectator. October 4, 1848.

———. February 14, 1849, 2. Letter from an anonymous correspondent presumably from the gold fields dated November 24, 1848.

———. March 14, 1849. "Ho! For California."

———. April 11, 1849. "For California."

Owen, J. R. N. April 15, 1861. Visalia, CA. "Report of J. R. N. Owen to Hon. S. Mowry of A Reconnaissance Along the Proposed Boundary Line of the State of California." Sacramento: California State Land Commission Files.

Palmer, T. S. February 8, 1894. Washington, DC. Letter to L. U. Richards. JA780.

———. April 9, 1894. Washington, DC. Letter to Col. John B. Colton. JA766.

———. September 23, 1894. Washington, DC. Letter to John B. Colton. JA769.

———. March 9, 1910. Washington, DC. Letter to Col. John B. Colton. JA777.

———. 1980. *Place Names of the Death Valley Region in California and Nevada*. Edited by Daniel Cronkhite. Morongo Valley, CA: Sage Brush Press. An abbreviated version appeared as an appendix in Palmer, "List of Locations Visited by the Death Valley Expedition." In *North American Fauna, No. 7*, 361–84. Washington, DC: GPO, 1983.

———. 2006. *Chronology and Names of the Death Valley Region in California, 1849–1949*. Rockville, MD: Wildside Press. A facsimile of *Chronology of Death Valley, California, 1849–1949 and Place Names of the Death Valley Region in California and Nevada, 1845–1947*.

Perkins, Arthur B. 1954. "The Story of Our Valley: Rancho San Francisco." http://www.scvhistory.com/scvhistory/signal/perkins/part02.html.

———. 1957. "Rancho San Francisco: A Study of a California Land Grant." *Historical Society of Southern California, Quarterly* 39 (2): 99–126.

Pioneer (San Jose, CA). March 21 and 28, 1877. "Biographical Sketches: W. L. Manly—How He Crossed the Plains—Description of an Adventurous Journey—Death Valley Experience."

Pippin, Lonnie C. 1986. "An Overview of Cultural Resources on Pahute and Rainier Mesas on the Nevada Test Site, Nye County, Nevada." Technical Report no. 45. Desert Research Institute, University of Nevada System, Social Sciences Center. Prepared for USDE, Las Vegas, Nevada Operations Office.

Plummer. 1892. Obituary. *Stark Co. News*. N.d. after June 22, 1892. JCHL, Scrapbook 1.

Post, Frederick J. 1977. "The Microbial Ecology of the Great Salt Lake." *Microbial Ecology* 3: 143–65.

Pratt, Addison. 1990. *The Journals of Addison Pratt*. Edited by S. George Ellsworth. Salt Lake City: University of Utah Press.

Pratt, Orson, et al. [1947]. *Exodus of Modern Israel*. Compiled by N. B. Lundwall. Independence, MO: Zion's Printing and Publishing.

Price, R. C. December 2, 1889. Knoxville. Letter to Friend John B. [Colton]. JA793.

Purdy, John H. April 18, 1850. San Francisco. Letter to Dear Brother. Correspondence, 1849–1851, MSS 862. Harold B. Lee Library, Brigham Young University, Provo, UT.

Putnam, Daniel. 1899. *A History of the Michigan State Normal School: At Ypsilanti, Michigan, 1849–1899*. Ypsilanti: Michigan State Normal College.

Quade, Jack, and Joe Tingley. 2006. "Wahmonie." *Boomtown History: Centennial Celebration of Nye*

County & Death Valley Area Mining Camps, edited by Jean Johnson, 7–17. Amargosa, NV: Nevada Boom Town History Event.

Quaife, Milo Milton. See Carson, Kit.

Rasmussen, Louis J. 1994. *California Wagon Train Lists*. Vol. 1. *April 5, 1849 to October 20, 1852*. Coloma, CA: San Francisco Historic Records.

Record-Union (Sacramento). April 1898. "Jayhawkers of '49: Rounding Up the Stragglers of the Party of Argonauts." JCHL, Scrapbook 2.

Republican Register. March 5, 1887. "The Jayhawkers." JCHL, Scrapbook 1.

[*Republican Register*] (handwritten note) (Galesburg, IL). Could be *Knox Republican*. February 5, 1895. "The Jayhawkers Meet: Terrible Experiences in Death Valley." JCHL, Scrapbook 2, 142.

Ressler, Theodore C. 1964. *Trails Divided: A Dissertation on the Overland Journey of Iowa "Forty-Niners" of the "Sacramento Mining Company. Prepared from Original Sources and Compared with other Contemporary Accounts*." Williamsburg, IA: privately printed. Ressler attributed this log to David Switzer, but it is now known to be written by Welch, 255–74. Ressler mimeographed 110 copies of this work. Copy in Johnson files.

Richards, Luther A. January 4, 1879. Galesburg, IL. Letter to L. Dow Stephens. JA841.

———. August 28, 1889. Mascot, NE. Letter to Friend Colton, JA822.

———. January 10, 1894. Beaver City, NE. Invitation to Dear Comrade. JA844.

Ricketts, Norma Baldwin. 1996. *The Mormon Battalion: U.S. Army of the West, 1846–1848*. Logan: Utah State University Press.

Rocky Mountain News (Denver). October 30, 1919.

Rogers, John Haney. April 26, 1894. In *Merced (CA) Star*.

———. See Johnson and Johnson 1999.

Rood, Wm. B. July 26, 1859. Tucson, AZ. Letter to John [Colton]. JA849.

———. Circa 1860. Letter to Colton, [First page(s) missing.] JA852.

———. April 28, 1867. La Paz, AZ. Letter to Mr. John Colton. JA853.

Royce, Sarah. 1939. *A Frontier Lady: Recollections of the Gold Rush and Early California*. Edited by Ralph Henry Gabriel. New Haven, CT: Yale University Press.

Runner, Meredith. 1995. *"The '49 Gold Rush, Jayhawker Wagon #9, and the Bartholomew Bear."* Typescript. In Johnson files.

San Francisco Call. February 9, 1913. "Jayhawkers at Dinner Once More: Of Original Large Desert Party Only Four Are Living."

San Francisco Chronicle. February 15, 1903. "Story of the Jayhawkers: The Pioneer Party That Discovered Death Valley and Its Nitre Deposits," 3.

San Joaquin County Death Certificates Index. Stockton, CA.

San Jose Daily Mercury. November 16, 1903. "Thomas Shannon's Journey of Peril." Obituary.

San Jose Mercury Herald February 5, 1918.

Santiago [James Martineau]. 1890. "Seeking a Refuge in the Desert—II." *Contributor* 11 (8): 296–300. Salt Lake City: Deseret News.

Seeley, David. 1849. "David Seeley Sketch," taken from "Biographical Sketch of David Seeley." Bancroft Library, Berkeley. In *Journals of Forty-Niners*, edited by LeRoy R. Hafen and Ann W. Hafen, 296–98. Lincoln: University of Nebraska Press, 1954.

Service, Robert. *Songs of a Sourdough*. 1907.

Seymour, H. M. June 5, 1849. Letter sent from Fort Laramie to J. S. Swezy. *Oquawka (IL) Spectator*, July 18, 1849.

Shallat, Todd. 2013. "Machines in Desolation: Images of Technology in the Great Basin of the American West." *Historical Geography: An Annual Journal of Research, Commentary and Reviews* (University of New Mexico) 41: 156–90.

Shannon, Ralph W. (Thomas's son). November 26, 1903. Letter to John B. Colton, JA860.

Shannon, Thomas. March 25, 1854. Marysville, CA. Letter to Friend Colton. JA863.

———. January 18, 1879. Los Gatos, CA. Letter to Dear Deacon [Richards]. JA875.

———. December 9, 1881. Los Gatos, CA. Letter to Capt Asa Haynes. JA869.

———. January 15, 1893. Letter to Colton. JA865.

———. January 17, 1894. Letter to L. A. Richards and all old Jayhawkers. JA878.

———. January 21, 1895. Los Gatos, CA. Letter to A. C. Clay and fellow Jay Hawkers. JA862.

———. January 22, 1897. Mount Seir Los Gatos, CA. Letter to Dear Sir [Colton]. JA867.

———. October 1, 1897. Los Gatos, CA. Letter to J. B. Colton. JA866(1).

———. September 4, 1898. Los Gatos, CA. Letter to J. B. Colton. JA866(2).

———. January 25, 1899. Los Gatos, CA. Letter to John B. Colton. JA868.

———. January 10, 1900. Los Gatos, CA. Letter to old Jayhawkers. JA881.

———. January 10, 1902. Soldiers Home, Los Angeles. Letter to The Jayhawkers of '49. JA882.

———. February 1, 1903. Soldiers Home, Los Angeles County, "permanent Address Los Gatos." Letter to Jayhawkers of '49. JA883.

———. November 16, 1903. Obituary. "Thomas Shannon's Journey of Peril." *San Jose Daily Mercury.*

Shumway, Burgess McK. 1988. *California Ranchos: Patented Private Land Grants Listed by County.* San Bernardino, CA: Borgo Press.

Smith, Wallace E. 1977. *This Land Was Ours: The del Valles and Camulos.* Ventura, CA: Ventura County Historical Society.

Spears, John R. 1892. *Illustrated Sketches of Death Valley and Other Borax Deserts of the Pacific Coast.* Chicago: Rand McNally.

Spokane Daily Chronicle. January 4, 1946. "Paul De Laney, Attorney, Dies."

Statutes at Large. 4 Stat. 729. 1846. "An Act to regulate Trade and intercourse with the Indian tribes, and to preserve peace on the frontiers." June 30, 1834. In vol. 4 of *The Public Statutes at Large of the United States of America.* Boston: Charles C. Little and James Brown.

Steiner, Harold. 1999. *The Old Spanish Trail Across the Mojave Desert.* Las Vegas: Haldor.

Stephens, Lorenzo Dow. January 21, 1880. A report of the 1880 Jayhawker reunion, JCHL, Scrapbook 1.

———. January 25, 1881. San Jose, CA. Letter to Capt Asa Haynes and Brother Jayhawkers. JA970.

———. March 16, 1884. San Jose, CA. Letter to J. B. Colton. JA898.

———. January 31, 1886. Strawberry Valley, CA. Letter to Luther A. Richards. JA977.

———. January 20, 1887. San Jose, CA. Letter to Friend Mecum and Brother Jayhawkers. JA972.

———. January 24, 1892. San Jose, CA. Letter to Chas B. Mecum and Brother Jayhawkers of '49. JA973.

———. January 22, 1893. San Jose, CA. Letter to J. B. Colton & Brother Jayhawkers. JA899.

———. January 20, 1894. San Jose, CA. Letter to Dear Friend Richards and Brother Jay Hawkers. JA987.

———. January 28, 1895. San Jose, CA. Letter to Brother Jayhawkers. JA896.

———. January 16, 1896. San Jose, CA. Letter to Mr. Chas Mecum and Brother Jayhawkers of '49. JA974.

———. January 30, 1899. San Jose, CA. Letter to John B. Colton & Brother Jayhawkers, JA902.

———. January 16, 1901. 565 Martin Ave. San Jose, CA. Letter to John B. Colton. JA904.

———. January 20, 1902. San Jose, CA. Letter to The Jayhawkers of '49. JA981.

———. February 8, 1903. 565 Martin Ave. and 371 W. Santa Clara St. [business]. Letter to J. B. Colton. JA907.

———. [Handwritten note: February 4], 1905. San Jose, CA. Letter to Mr. Chas B. Meacum and Comrads Jayhawkers of '49. JA975.

———. July 12, 1908. San Jose, CA. Letter to John B. Colton. JA917.

———. January 24, 1909. San Jose, CA. Letter to Dear friend and Comrad [Colton]. JA918.

———. 1916. *Life Sketches of a Jayhawker of '49: Actual Experiences of a Pioneer Told by Himself in His Own Way.* San Jose, CA: Nolta Bros.

Steward, Julian H. 1938. *Basin-Plateau Aboriginal Sociopolitical Groups*. Smithsonian Institution, Bureau of American Ethnology Bulletin 120. Washington, DC: GPO. Reprint, Salt Lake City: University of Utah Press, 1970.

Stoffle, Richard, David B. Halmo, John E. Olmsted, and Michael J. Evans. 1990. *Native American Cultural Resources Studies at Yucca Mountain, Nevada*. Ann Arbor: Institute for Social Research, University of Michigan.

Stott, Clifford L. 1984. *Search for Sanctuary: Brigham Young and the White Mountain Expedition*. Salt Lake City: University of Utah Press.

Stover, Jacob Y. 1937. "History of the Sacramento Mining Company of 1849: Written by One of Its Number." Edited by John Walton Caughey. *Pacific Historical Review* 6 (2): 166–81. Part of this article is in LeRoy R. Hafen and Ann W. Hafen, eds., *Journals of Forty-Niners*, 274–91.

Stretch, Richard H. 1867. "Journal of Explorations in Southern Nevada in the Spring of 1866 by His Excellency Governor Blasdel of Nevada." Appendix E in *Annual Report of the State Mineralogist of the State of Nevada for 1866*, 141–47. Carson City, NV: Joseph E. Eckley, State Printer.

Sutak, Tom. 2010. "Jefferson Hunt: California's First Mormon Politician." *Journal of Mormon History* 36 (3): 82–117.

———. 2012. *Into the Jaws of Hell: Jefferson Hunt: The Death Valley '49ers Wagon Train & His Adventures in California, 1846–1857*. Danville, CA: Pine Park.

———. March 21, 2016. Personal communication to Johnson and Johnson.

Taylor, Robert B. August 5, 1854. Stockton, CA. Letter to My Dear John [Colton]. JA989.

Thomason, Jackson. 1978. *From Mississippi to California: Jackson Thomason's 1849 Overland Journal*. Edited by Michael D. Heaston. Austin, TX: Jenkins.

[Thompson, Frank.] May 2, 1890. "The Place They Came To." *Ventura (CA) Free Press*.

Thompson, G. H. 1874. "Plat of the Rancho San Francisco finally confirmed to Jacoba Feliz et al. Surveyed under instructions from the U.S. Surveyor General. By G. H. Thompson, dep. sur. June 1874." Bureau of Land Management, Folsom, CA. Docket 399, F1 & F2, map in two parts. Copy in Johnson files.

TOPO! 2009. *TOPO! Outdoor Recreation Mapping Software: California, Nevada, Utah*. National Geographic Maps, edition 4.5.4.

Tschanz, C. M., and E. H. Pampeyan. 1970. *Geology and Mineral Deposits of Lincoln County, Nevada*. Nevada Bureau of Mines and Geology, Bulletin 73. Mackay School of Mines. Reno: University of Nevada Press.

United States of America. February 12, 1875. "Patent." Patent for Rancho San Francisco issued to Jacoba Felis, Ygnacio del Valle, Maria del Valle, Magdelina del Valle, José Antonio del Valle, José Ygnacio del Valle, and Concepcion del Valle. Filed March 18, 1875, in book 1 of Patents, Page 514, office of the County Recorder, Los Angeles County, CA. Signed by President U.S. Grant. Typescript copy in Johnson files.

U.S. Census of California, 1850. See Newmark.

U.S. Census Records. Various years. Online sources.

[Van Dorn, Aaron]. 1861. "Eastern Boundary Sketches. By One of the Exploring Party of the Late U.S. Boundary Commission." *Sacramento Daily Union*, June 25, 29, July 9, 11, 13, 31, August 7 & 10. Reprinted in Woodward 1961. These articles were not bylined; we identified the author by deduction.

Vincent, J. M. 1911. *Historical Research: An Outline of Theory and Practice*. New York: Henry Holt.

W. [Wood, Newton]. May 16, 1849. *Oquawka (IL) Spectator*.

———. June 5, 1849. Fort Laramie. Letter to Colonel. "For the Oquawka Spectator." *Oquawka (IL) Spectator*. August 8, 1849.

Walker, Ardis M. 1961. *Freeman Junction: The Escape Route of the Mississippians and Georgians from Death Valley in 1849*. San Bernardino, CA: Inland Printing and Engraving.

Wasley, James. October 21, 1849. "California Letter." In *Wisconsin Tribune* (Mineral Point), March 1, 1850.

Webster, Kimball. 1917. *The Gold Seekers of '49: A Personal Narrative of the Overland Trail and Adventures in California and Oregon from 1849 to 1854*. Manchester: N. H. Standard Book.

Weight, Harold, and Lucile Weight. 1959. *Wm. B. Rood, Death Valley 49er, Arizona Pioneer, Apache Fighter, River Ranchero, and The Treasure of Ruined "Las Yumas."* Twentynine Palms, CA: Calico Press.

Welch, Adonijah Strong. September 30 to October 14 and October 20 to November 5, 1849. "Journal" (or Log). Unpublished. Copy in Bancroft Library, Berkeley, CA. Called "The Switzer 'Log'" in Ressler, *Trails Divided*.

West, J. L. February 1, 1897. Phillipsburg, MT. Letter to Mr. John B. Colton. JA1013.

Wheat, Carl I. 1939a. "The Forty-Niners in Death Valley: (A Tentative Census)." *Historical Society of Southern California Quarterly* 21 (4): 102–17.

———. 1939b. "Hungry Bill Talks." *Westways* 31 (5): 18–19.

———. 1939c. "Pioneer Visitors to Death Valley After the Forty-Niners." *California Historical Society Quarterly* 18 (September): 195–216.

———. 1939d. "Trailing the Forty-Niners Through Death Valley. *Sierra Club Bulletin* 24 (3): 74–108.

———. 1940. "The Jayhawkers at the Missouri: A Remarkable Discovery." *Historical Society of Southern California Quarterly* 22 (3): 103–8.

———. 1956. "Mapping the American West: A Bibliographical Summary." *Papers of the Bibliographical Society of America* (New York), Far Western Issue, 50 (1): 1–16.

———. 1959. *Mapping the Transmississippi West.* Vol. 3, *From the Pacific Surveys to the Onset of the Civil War, 1846–1845.* San Francisco: Institute of Historical Cartography.

———. 1960. *Mapping the Transmississippi West.* Vol. 4, *From the Pacific Railroad Surveys to the Onset of the Civil War, 1855–1860.* San Francisco: Institute of Historical Cartography.

Wheeler, George M. [1875?]. Base map for Atlas Sheet 65 (D). Hand drawn. National Archives, Washington, DC: U.S. 542-17. Copy in Johnson files.

Wheeler, George M. 1877. *Geographical Explorations and Surveys West of the 100th Meridian, Topographical Atlas, Part of Eastern California, Atlas Sheet 65 (D).* Scale 1/253440.

White, William Wellington. 1927. "William Wellington White: An Autobiography." *Quarterly of the Society of California Pioneers* 4 (4): 202–16.

Wiley, Nancy J. Haynes. 1937. "Story of the Jay-Hawkers of '49 from Accounts of Captain Asa Haynes." Presumably handwritten. "Typed by Nancy Cramer [Nancy Wiley's grand daughter], June 28, 1937." Knox College Archives, Galesburg, IL. 987-3-1. 12 pp.

Wiley, W. H. [Asa Haynes's son-in-law]. January 14, 1889. De Long, IL. Letter to Mr. Clay. JA1028.

Wilkins, James F. 1968. *An Artist on the Overland Trail: The 1849 Diary and Sketches of James F. Wilkins.* Edited by John Francis McDermott. San Marino, CA: Huntington Library.

Woodward, Arthur, ed. 1949. *The Jayhawkers' Oath and Other Sketches.* Los Angeles: Warren F. Lewis.

———. 1961. *Camels and Surveyors in Death Valley.* Palm Desert, CA: Death Valley '49ers. When Woodward published this booklet, he did not know Aaron Van Dorn was the author of this series of articles that appeared in the *Sacramento Daily Union* between June and August 1861.

Young, Lewis. July 17, 1897. Rutland, IL. Letter to Mr. John B. Colton. JA1037.

———. November 26, 1900. Santa Fe. Letter to Mr. John B. Colton. JA1039.

Young, Sheldon. 1849a. March 18, 1849–December 15, 1850. "Diaries of his trip to California, 1849–1850." San Marino, CA: Huntington Library, Western Historical Manuscripts, mssHM 75663–75664. The surviving pages of the original Sheldon Young diary donated by his great, great grandchildren, David Duerr and Gretchen Forsyth, 2014.

———. 1849b. "Sheldon Youngs [sic] Log, 1849, Joliet Illinois to Rancho San Francisquito, California." Typescript of original diary, 1897. JA555. Title was added by John Colton.

———. 1849c. Diary. Electronic scans made by David R. Duerr, Sheldon Young's great-great-grandson. In Johnson files. 2012. See also Sheldon Young "Diaries of his trip to California, 1849–1850." San Marino, CA: Huntington Library, Western Historical Manuscripts, mssHM 75663–75664.

———. 2010. *The Lost Jayhawker Story: Sheldon Young with the Death Valley '49ers.* Edited by David R. Duerr. Indian River, FL: Indian River.

ABOUT THE AUTHOR

JEAN JOHNSON, with her husband, LeRoy, has been active in searching the trails and truths of early Death Valley history for the past forty years. She has been secretary-treasurer and editor of the proceedings for seven Death Valley history conferences and has written several papers for those conferences, plus producing three Nevada Boomtown Conference proceedings. Jean has served on the board of directors of the Death Valley '49ers, Inc., and was their publications chairman for several years. She has written several keepsakes for them, plus coauthoring with LeRoy *Julia, Death Valley's Youngest Victim; Escape from Death Valley;* and a well-footnoted and indexed edition of William Lewis Manly's *Death Valley in '49* for Santa Clara University. For fun she edited and compiled *Tall Tales of Death Valley*. Jean is also a cellist who has served as first chair of the Sacramento Symphony and Albuquerque and Bishop chamber orchestras plus numerous smaller performing groups. She also plays in a handbell choir and conducts another one. Her current musical interest is learning more about and performing Baroque music. Born in Colorado, most of her schooling was in the San Francisco Bay Area, including graduation from Mills College. She later took editing classes at the University of Minnesota. Rearing two sons, Eric and Mark, has been a major joy of her life. Jean and LeRoy currently reside in McMinnville, Oregon, but they make yearly trips to Death Valley. They are currently preparing for publication a book on the very fine California journals of young diarist William B. Lorton, who traveled from Illinois to the California goldfields in 1849.

INDEX